CARE-S

Computer Aided Rehabilitation of Sewer and Storm Water Networks

CARE-S

Computer Aided Rehabilitation of Sewer and Storm Water Networks

Edited by

Sveinung Saegrov

Publishing

LONDON • SEATTLE

Published by IWA Publishing, Alliance House, 12 Caxton Street, London SW1H 0QS, UK

Telephone: +44 (0) 20 7654 5500; Fax: +44 (0) 20 7654 5555; Email: publications@iwap.co.uk
Web: www.iwapublishing.com

First published 2006

British Library Cataloguing in Publication Data
A CIP catalogue record for this book is available from the British Library

Library of Congress Cataloging- in-Publication Data
A catalog record for this book is available from the Library of Congress

ISBN: 1843391155

ISBN13: 9781843391159

Contents

Foreword

CARE-S was a project under the Fifth Framework Program of the European Commission from 2002 to 2005. The project had a budget of 5 million Euro. It included fifteen partners from ten countries in Europe and Australia. Additionally twenty cities acted as end-users. The main CARE-S outcome is a comprehensive decision support system (DSS) for the wastewater network management. The CARE-S DSS is meant for rehabilitation planning of sewer and storm water pipes.

This report comprises a concentrated description of the methods used, the tools and software produced and the results from using it. The entire CARE-S documentation consists of 21 software, 31 deliverables and a large number of reports and publications. Most of the information is public and can be accessed from the CARE-s web (http://care-s.unife.it). A comprehensive testing program has clearly demonstrated that this system has a potential to support a proactive rehabilitation strategy.

The water network planner will hopefully see the CARE-S logo and name in many years from now. It is our ambition that the CARE-S system will be a leading instrument for the water industry in upgrading their wastewater networks and for general management of the networks. We intend to establish an organisation to link the international companies and institutes working with CARE-S. At least from the beginning the end users will meet the current CARE-S partners as the CARE-S service providers.

We would like to thank the partners and the end users who have participated in the CARE-S project. Without the end users, it would not have been possible to complete this project in practice. Without the scientific competence of the partners, CARE-S could never have been realized. CARE-S has not been a straightforward exercise. It relies heavily on many years of research that includes many doctoral degrees. The CARE-S project has built on existing scientific work and developed new elements, to provide a complete system for water network rehabilitation.

We would also like to thank the European Commission and the administration of the Fifth Framework Programme that made CARE-S possible by its financial support. Additionally, we thank the national agencies that provided the remaining resources for the fulfilment of CARE-S.

Trondheim December 2005
Sveinung Sægrov
CARE-S Co-ordinator

Summary

Sewer and storm water systems in cities worldwide suffer from insufficient capacity, construction failures and pipe deterioration. The consequences are structural damage and local floods leading to inflow of water into basements, traffic disturbances, street and surface erosion, and pollution of local receiving waters.

The reasons for the problems are the combined effects of ageing infrastructure, urbanisation and climate change. European cities spend in the order of 5 billion € per year for wastewater network rehabilitation. This amount will increase significantly over the coming decades, due to network ageing.

Today, the rationale behind rehabilitation decisions is unclear. Decisions are often made ad hoc and, in the best case, are based on practical experience after failures appear (reactive approach). This is the background for a project in the 5th Framework Program in EU, Computer Aided Rehabilitation of Sewers and Storm water networks (CARE-S). The project objective is to establish a rational framework for sewer network rehabilitation decision-making. CARE-S aims to improve the structural and functional reliability of the wastewater networks (risk of in- and exfiltration, collapse and blockage due to pipe deterioration, hydraulic overloading resulting in flooding and/or receiving water pollution). This is combined with the analysis and, where necessary, upgrading of the network capacity to take care of runoff from rainstorms and snow melt in urban areas. The ultimate product is a Decision Support System (DSS) that enables municipal engineers to establish and maintain effective management of their sewer networks. In other words: *Rehabilitate the right sewer at the right time by using the right rehabilitation technique at a minimum total cost, before serious failures occur (pro-active approach).*

Some analytical tools to assess the technical or functional state of sewers or the needs for rehabilitation are available or are under development. Usually, though, they do not take into account all aspects of rehabilitation decisions, and they only have been applied for a limited number of wastewater services.

The objectives of CARE-S are to improve these tools, link them and make them usable for the formulation of a rehabilitation strategy. The DSS includes the following elements (Figure 1):

- a tool generating Performance Indicators (PI) that are relevant for rehabilitation decisions, including analytical and statistical procedures to assess and forecast some of the PIs
- a procedure to define the socio-economic and environmental risks of malfunctioning sewer systems
- a database to be used for choosing an appropriate rehabilitation technology
- several tools that allow to assess the hydraulic, environmental and structural conditions of the network including their change over time.
- a tool to define the best long-term strategy for rehabilitation investments
- a multi-criteria decision (MCD) tool supporting the choice of high priority rehab projects
- a software package, called "Sewer Rehab Manager" that will enable consultants and wastewater service providers to use the above products according to their individual needs and available data base.

The project has been carried out as a joint European initiative supported by the European Commission under the Fifth Framework Programme for Research and Development, and had a budget of 5 million Euro. 14 European and 1 Australian partner are involved, together with 20 cities ("end-users") representing all parts of Europe. The partners were: SINTEF, Norway (co-ordinator), Bologna university, Italy, Brno University, Czech republic, Budapest university, Hungary, Cemagref, France, CLABSA, Spain, CSIRO, Australia, TU Dresden, Germany, Ferrara university, Italy, LNEC, Portugal, Norwegian university of Science and Technology, Palermo university, Italy, WRc, United Kingdom, Aalborg university, Denmark Additionally, a scientist from Czech Technical University has participated.

The project CARE-S started 1. October 2002 and finished 30. September 2005. The final stage of project work included a comprehensive testing and validation in collaborating cities.

A brief explanation of the system is given in the following pages.

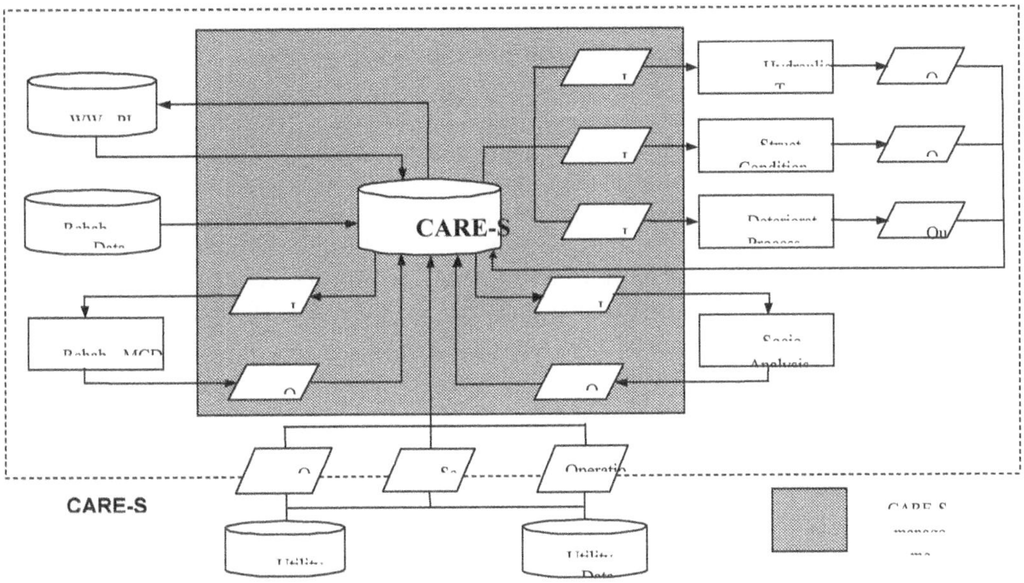

Figure 1. The general architecture of the CARE-S Sewer Rehab Manager

Construction of a control panel of Performance Indicators (PI) for rehabilitation

A PI (Performance Indicator) tool addressing information relevant for sewer and storm water network rehabilitation has been developed. It is based upon comprehensive and quality-controlled information from 32 project partners and end-users. The final listing includes 41 PI in total. Additionally, a large number of Utility Information (UI) and External Information (EI) have been included in the system.

The general criteria for selecting PI for wastewater rehabilitation were among others:

- to represent relevant rehabilitation aspects of wastewater undertakings performance, allowing for a global representation of the system by a reduced number of performance indicators
- to be clearly defined with a concise meaning and a unique interpretation for each indicator
- to be auditable, which is specially important when the performance indicators are to be used by external bodies (i.e. regulators) that may check the results reported.

A questionnaire to verify 35 pre-selected PI was answered by 13 Partners and 20 End-Users. The receivers were asked to rank the PI regarding importance and access. Based on the answers, the list of PI was extended to 41.

The questionnaire revealed that partners tend to overestimate the importance of PI in comparison with end-users. In general the Partners also tend to be more "optimistic" than End-Users concerning the PI assessment.

PI relies on qualified definitions. Therefore, a glossary of rehab terms has been developed. It includes mainly a set of definitions referred in European Standards, with some additional definitions from WRc Rehabilitation manual.

Description and validation of technical tools for structural condition

The structural performance package comprises network level models as well as detailed pipe-specific models of corrosion, load-bearing capacity, blockages and in/ex-filtration.

The structural condition of a pipe depends on material, construction practices, external load and wastewater characteristics. Previous research has shown that pipes laid within certain time periods have been structurally under-designed, and that construction practices within some time periods have not been appropriate, thus leading to particularly frequent failures. The formation of hydrogen sulphide that occurs in some wastewater networks from anaerobic decomposition of organic mater compounds or from reduction of mineral sulphides will lead to a microbiological deterioration and loss of strength for concrete sewers, and is a very important reason for rehabilitation actions. Other reasons are root intrusion and substantial in- and ex-filtration through fissures and leaky joints.

Structural analysis of sewers is normally based on results from CCTV inspection. In Europe there are several systems currently being used for the classification of individual data from such investigations. A standard European code for the description of sewer damages has been developed and recommended to the member

States. In general, the condition of sewers is classified according to its most severe damage or overall condition to determine the urgency of rehabilitation and calculate the cost of rehabilitation.

In CARE-S, different models for the classification and assessment of the sewer condition is analyzed and tested with CCTV data together with other types of data, provided by end-users. Network aggregated models calculate the distribution of condition defined by classes. Pipe specific structural models deal with issues important for the hydraulic performance and structural condition on a local (pipe based) level. They comprise structural failures, strength reduction due to hydrogen sulphide attack or external corrosion, pipe blockage and in/exfiltration of water. The models are developed or improved and calibrated with data from the end-user sewer networks. The models allow a forecast of the year in which a sewer will fail or enter a critical class of condition, thus determining the next inspection date or rehabilitation measure. An overview of models is presented in Figure 2.

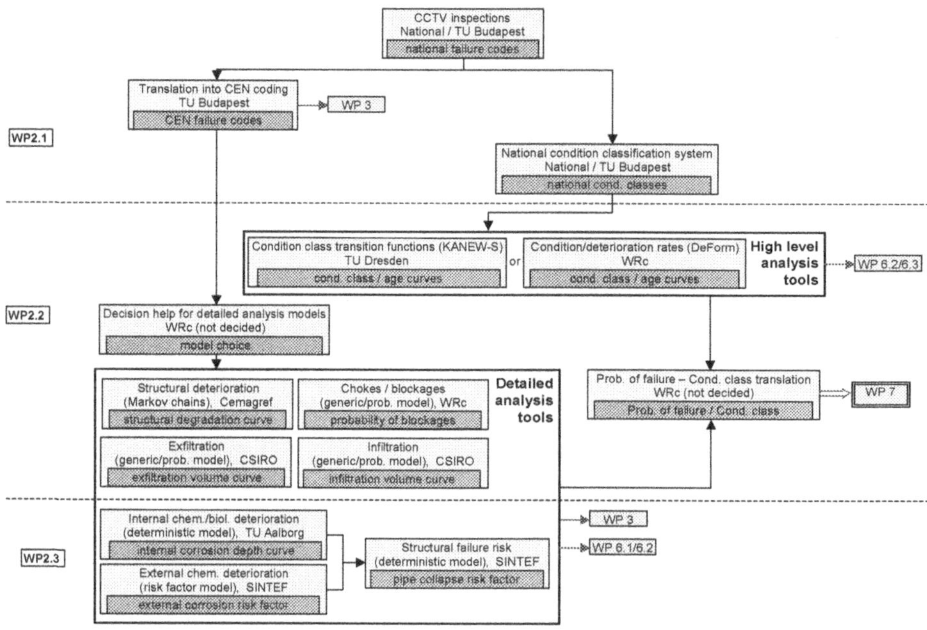

Figure 2 Overview of models for structural assessment of wastewater networks

Description and validation of hydraulic performance

Structural decline of sewers is described by CEN-code classes (e.g. obstacles, roots, displacement of joints, intruding pipes). Their hydraulic effects are modelled using 3-D CFD software (FLUENT) and "translated" into a simplified 1-D description for application in hydraulic models.

A study on best practise on wastewater modelling has demonstrated that single models have reached high levels of reliability. Still further progress is needed on model integration in order to achieve a global control on urban water analysis. It has be decided to include three 1D models as basis for routines developed within CARE-S, namely MOUSE, INFOWORKS and SWMM 4.4.

A method to estimate temporal decline of hydraulic performance has been developed. CCTV observations and a 3D CFD software (FLUENT) have been used for parameterization and validation of 1D hydraulic model (MOUSE, InfoWorks, SWMM). Test models at catchments of Palermo, Reggio Emilia and Oslo have been prepared for the testing of the routine that is developed in this work package, and the results of testing were successful.

A procedure has been developed for assessing compliance of the sewer system with the environmental standard/target/threshold specified by the end-user (effects of rehabilitation works on ground and surface water) using conceptual infiltration/ex-filtration models.

The results on hydraulic performance are also connected to analysis of changes in future structural performance that may affect the hydraulic capacity, and socio-economic consequences. This integration makes it possible to analyse future hydraulic performance and the consequence related to flooding in sensitive areas.

Rehabilitation technology information system

An extensive rehabilitation and repair technology database including available techniques, their advantages, disadvantages and application possibilities has been developed. The database also gives an estimate of technology costs.

Rehabilitation of sewer and storm water networks includes system improvements (detention basins, separation of storm water etc) as well as renovation technology. During the last 20 years, there has been an extensive development of technologies for sewer and storm water rehabilitation. Today, a large variety of methods exist to meet problems in small sized as well as large sized pipelines. Several cities use renovation as their first priority measure against structural and hydraulic decline.

The repair or renovation technology to be chosen obviously depends on local conditions, such as the type of the problem, the size of the pipe etc. The rehabilitation costs are linked to the applied method and technology. By using the database, the End User is invited to analyze which measures are appropriate to meet his various kinds of problems. A list of measures and problems they should apply to is presented, considering their potential for pipe failure in one hand, and their characteristics leading to more or less social disturbance and annoyances on the other hand. The content of the rehabilitation technology information system is, briefly summarized:

- General survey of available techniques and contractors. Analysis of experiences and results.
- Complete chart for the variety of methods in rehabilitation of sewers. Current state-of-the-art, including rates, range of applicability, limits and restrictions.
- Cost of rehabilitation linked to the applied technology.
- Alternatives for rehabilitation: redesign of sewers, operational methods.
- Criteria for choice making of suitable techniques related to sewer problems.

Also, attention has been brought to the innovative question of establishing the explicit links between the rehabilitation techniques (with or without trench) and the other aspects of wastewater rehabilitation, such as disturbances to the citizens, hydraulic benefits, mechanical benefits and durability, and economical issues.

Socio-economic consequences

Socio-economic consequences including rehabilitation impact on socio-economic cost and quality of life, public acceptance and communication with public is investigated.

Rehabilitation impact on socio-economic costs

Very little quantitative information is available on the direct and indirect costs of wastewater network failures. Information is also scarce and scattered on social costs of rehabilitation works.

CARE-S will help addressing the socio-economic ("indirect" or "external") costs linked to rehabilitation decision: impacts of failures and impacts of rehabilitation works to third parties. The definition of "social costs" considered by the project is: the costs incurred by society as a result of sewerage works and for which utilities or companies have no direct responsibility out of possible compensation. The operational objective of this is to provide guidance and methods for assessing criteria for rehab planning representing social costs. This can be divided in two parts:

The first deals with criteria on external socio-economic impacts of rehabilitation works for comparing a limited set of technologies when considering a given single pipe. The technologies are described through outputs of the rehab technology information system. It provides criteria exclusively for selection of technology to specific projects.

The second deals with criteria on external socio-economic impacts of network failures for comparing various rehabilitation projects, defined each at pipe level. The failures or potential reduction of failure risks thanks to rehabilitation are related to each single pipe (the criteria have to consider the impacts at this scale), and are described through results of the structural condition and hydraulic performance modules.

This task includes definition of a conceptual framework accounting for the links between technical, economic and social aspects, and bases on both synthesis of literature (methods, results of studies) and analysis of real world data (claims, compensations, failure events).

In connection with this analysis of field data, there is also an analysis of utility internal costs of social accompanying tasks linked to failures and works (management of claims...): These costs are on the border between internal and external social costs, they are usually not addressed as such or specifically quantified by utilities.

Rehabilitation impact on social life quality

This work has provided a systematisation of this issue:

- What the different users perceive as being the costs, e.g. economic, social or personal, of rehabilitation works.
- What are the public levels of awareness to the importance of maintenance actions, repair and rehabilitation of sewer and storm water networks? This identifies a need to evaluate the extent to which individuals are aware of non-intervention possible risks and consequences related to network failures
- What are the levels of public tolerance to a set of interventions in their work or residential area?

Research on these three parameters has been developed through the use of public inquiry methodologies (exploratory interviews, focus groups and questionnaire) in selected areas (Oeiras/Amadora and Nantes).

Public acceptance and communication with public

Based on the interviews, focus groups and questionnaire surveys, this task aimed to provide decision makers with a better understanding of public perception and acceptance concerning sewer systems failures and rehabilitation (needs and impacts). This is meant to enable better communication with the public including prior explanations on projects and public participation.

Multi-criteria decision support

Defining an appropriate long-tem rehabilitation strategy

Investments in sewer rehabilitation must be based on inspection and evaluation of sewer conditions with respect to the severity of sewer damage and to environmental risks. Key issues of strategic or long-term rehabilitation planning of a sewer network are:

- Forecasting the future condition of sewer pipes by deterioration models
- Determination of deterioration rates and the associated rehabilitation needs for
  (a) maintaining the actual condition of the network in the long term
  (b) reaching an improved level of service within given time horizon
- Forecasting performance indicators of the network for different strategies (scenarios)
- Evaluation of different scenarios, and choosing the best rehabilitation strategy

Within CARE-S, condition forecasts are performed using the CARE-S WP2 GompitZ model. Transition functions from one into the next poorer condition class are empirically derived from a sample of inspected sewer pipes. The transition probabilities are used to forecast the future condition of sewers. By the same procedure, transition probabilities have been calibrated for sub-samples of different types of sewers. With these probability functions, the most probable date of entering a critical condition class can be forecasted from pipe characteristics, such as material, period of construction, location, use for waste and/or storm water, profile, diameter, etc.

Setting priorities for rehabilitation projects

In general, the total cost of all viable rehabilitation projects exceed the available budget, thus not all sewers in critical condition can be rehabilitated immediately, and it becomes necessary to select those rehabilitation projects promising highest cost efficiency.

Rehabilitation projects in a sewer network are selected from structural, hydraulic, environmental and socio-economic criteria and the associated direct and indirect costs. External effects are considered by the risk associated with the deficiency of the sewers, for example the potential damage due to frequent flooding or the potential groundwater pollution due to an exfiltrating sewer. Supported by a geographical presentation of the network, the candidates for rehabilitation are selected, step-by-step, in an interactive elimination process. Within the procedure, which was successfully applied to the rehabilitation planning of water distribution networks, those candidates are eliminated which are not fulfilling specified thresholds. The remaining candidates have a bundle of attributes favourable for rehabilitation.

Choosing the right rehabilitation technology

The right/best sewer rehabilitation technique, in economic as well as technical terms, is chosen from a set of candidates fulfilling the requirements under specific local conditions. Direct rehab costs are systematically analysed and documented for a variety of open trenching and no dig rehabilitation technologies. Beyond these direct costs, the support system takes into account a multitude of other factors, which are usually collected by wastewater companies preparing a public tender on a specific rehab project. Although from a financial viewpoint, the waste water company would choose the lowest bid, it has also to consider external costs that are not charged directly to the waste water company, such as increased operating costs and travel times for road users.

Information on the new technologies is coming from the catalogue of rehab technologies, including more than 60 currently available rehabilitation technologies for repair, renovation and replacement.

Wastewater Network Rehabilitation Manager

The wastewater network rehabilitation manager is a procedure for integrating the CARE-S tools into a cohesive rehabilitation-planning package. The procedure is based on the integrated approach for rehabilitation planning described in EN752-5 and allows rehabilitation engineers to use the tools most appropriate to their planning needs and data availability.

The CARE-S Manager has been presented in a report with the following content:

(1) *A Generic Rehabilitation Planning Process*: a description of the general process that would be carried out when planning the rehabilitation of wastewater networks, regardless of the analysis tools available to the rehabilitation engineer. This set the scene for how the CARE-S procedure will improve the planning process.

(2) *Applying CARE-S to Rehabilitation Planning*: a description of how the CARE-S procedure may be applied to the generic rehabilitation planning process. This has included the identification and definition of methods or software tools that can be used under the CARE-S "umbrella" to develop rehabilitation plans.

User interface for the Rehabilitation Manager

The user interface can be defined as any element of the Rehabilitation Manager with which the user will interact. This would include a number of functions, such as:

- importing and manipulating data;
- accessing tools;
- editing data;
- displaying data and reporting the results of analysis, and;
- interactive guidance.

CARE-S Rehabilitation Manager.

The rehab manager software consists of a central MS Access 2000 (Visual Basic 6.0) database application with the following attributes:

- It provides a central storage area and reference point for essential CARE-S data;
- It accepts user-validated input data in a pre-defined format. Data source fields will be mapped to fields within the CARE-S database according to the specification of the declared input data source;
- It provides automatic conversion of data item units to standard CARE-S convention;
- It allows user interaction with a pre-determined range of tools under the CARE-S umbrella;

Acknowledgement

CARE-S was a research project supported by the European Commission under the Fifth Framework Program and contributing to the implementation of the key action "Sustainable Management and Quality of Water" within the Energy, Environment and Sustainable development program.

The partners were: SINTEF, Norway (co-ordinator), Bologna university, Italy, Brno University, Czech republic, Budapest university, Hungary, Cemagref, France, CLABSA, Spain, CSIRO, Australia, Dresden university, Germany, Ferrara university, Italy, LNEC, Portugal, Norwegian university of Science and Technology, Palermo university, Italy, WRc, United Kingdom, Aalborg university, Denmark. Additionally, a scientist from Czech university of Technology has participated.

REFERENCES

Matos R, Cardoso MA, Pinheira I, Almeida MC: Selection of a listing of Rehabilitation Indicators for Rehabilitation. CARE-S report D1 July 2003

Cardoso, MA, Matos R: Predictable performance Indicators, CARE-S report D2 April 2005

Knolmar M, Szabo G: Classification system based on visual inspection, CARE-S report D3, May 2003

Vollertsen J: Technical tools for structural condition , CARE-S report D4 June 2005

Vollertsen J: Technical tools fopr structural condition – deterioration processes, CARE-S report D5, June 2005

Freni G, Milina J, Maglionico M, DiFederico V: State for the art in Urban Drainage Modelling, CARE-S report D7, 2003

Milina J, Ugarelli R et al: Description and validation of hydraulic performance, CARE-S D8 April 2004

Schulz N, Krebs P: Environmental impact of rehabilitation, CARE-S report D9, July 2004

Montero C: Survey of operational options, CARE-S report D11, June 2005

Montero C, Villaneuva A, Hlavinek P, Hafskjold L: Wastewater rehabilitation technology survey. CARE-.S report D12, April 2004

Werey C, Torterotot JP, Sousa e Silva D, Kønig A, Peireira A, Montginoul M: Socio-economic impacts on socio-economic costs, CARE-S report D13, Oct 2004

Baur R, Herz R, Kropp I: Multi-criteria decision support methodologies, WP 6 conclusive report D19, September 2005

Hulance J, Orman N, Hurley R, Kowalski M: The CARE-S procedure, CARE-S D20, December 2003

Hulance J, Kowalski M, Taylor K, Hurley R: User interface for the CARE-S wastewater Rehab manager software, CARE-S report D21, April 2004.

1

Performance Indicators for Rehabilitation

Maria Adriana Cardoso, Rafaela Matos, Ilídia Pinheiro and Maria do Céu Almeida; LNEC

1.1 INTRODUCTION

CARE-S - *Computed Aided Rehabilitation of Sewer Networks* is a project funded by the 5th EU Framework Programme. It started in October 2002 and finished in October 2005.

The project goal was to establish a rational framework for decision-making support on sewer network rehabilitation. CARE-S aimed to develop an integrated and systematic methodology for rehabilitation. The final goal was to contribute to improve the structural and functional reliability of the sewer networks. The main deliverable is a Decision Support System (CARE-S prototype), to assist municipal engineers to establish and maintain effective management of their sewer networks; rehabilitate the right sewer at the right time by using the right rehabilitation technique at a minimum total cost before serious failures occur (pro-active approach).

The construction of a control panel of performance indicators (PIs) for rehabilitation is a component of the project (Work Package 1) developed by LNEC, aiming at: a) to define an appropriate set of rehabilitation performance indicators (PIs); b) to develop a software for easier PI assessment and field-testing with project end-users; c) to define a sub-set of "predictable" PIs for strategic long term rehabilitation planning.

1.2 PERFORMANCE INDICATORS FOR REHABILITATION

1.2.1 Background

Increasing diversity in the provision of services for water supply and wastewater disposal, together with greater accountability for the quality of those services, necessitate scrutiny using assessment systems that are consistent, transparent and auditable. Whilst a number of countries worldwide have well defined and regulated systems for evaluating the performance of water service providers, this is by no means global. It is within this context that the International Water Association (IWA) has developed systems of PIs for water supply as well as for wastewater services (Matos et al., 2003b).

The IWA PI system is intended to provide objective and comprehensive management tools for undertakings and other stakeholders involved in any aspect of water and wastewater service provision. The PIs have been developed independently from the level of economic development, or type of institutional system in which the undertaking operates, and allow globally diverse economic, demographic, cultural and climatic characteristics to be acknowledged. The PIs cover a wide range of activities comprising: management, personnel, financial, physical, operational, environmental and quality of service aspects.

Trends in PIs may show historically improving or deteriorating performance in time for remedial measures to be taken before major problems occur in service delivery. PIs require a large amount of data. Hence they can also provide the incentive for good monitoring, data recording and processing, and help decision makers focus scarce resource allocations into the key areas where data needs to be collected.

Some of the IWA PIs are appropriate for the purpose of system's rehabilitation but other more specific were required. Thus, having as the reference the IWA PIs framework, a tailored set of PIs has been selected within CARE-S, in collaboration with partners and end-users.

1.2.2 Requirements of performance indicators for rehabilitation

Within the CARE-S project, specific objectives of the rehabilitation PIs are:

- to help in the characterisation of the sewer network in terms of hydraulic, environmental and structural conditions, enhancing the evidence of weakness and vulnerability of certain areas (sectors, clusters, etc), thus demonstrating the need to intervene with rehabilitation planning and concrete measures;
- to provide key information to support a more pro-active approach in terms of rehabilitation, rather then reacting further to dysfunctions and crisis;
- to allow for assessment of the performance evolution, comparing the results achieved at a given time with those achieved in previous periods, or comparing the results actually achieved with preset targets or with reference values that have been agreed;
- to allow for more simple, objective and structured monitoring effects of rehabilitation management decisions (including socio-economical aspects);
- to facilitate the comparison of state condition versus performance versus investment at different locations or sub-systems inside a given undertaking, or compare, with similar undertakings, thus providing key information to promote future performance improvements.

The selection of a preliminary set of rehabilitation PIs based on the IWA system has taken into consideration the above-mentioned aspects keeping the PIs requirements as defined in Matos *et al.* (2003b), such as:

- to represent the relevant rehabilitation aspects of the wastewater undertaking's performance, allowing for a global representation of the system by a reduced number of performance indicators;
- to be clearly defined, with a concise meaning and a unique interpretation for each indicator;
- to be easy to understand, even by non-specialists, wherever possible;
- to refer to a well-defined period of time and a well defined geographical area;
- to be applicable to undertakings with different characteristics and stages of development.

1.3 Methodology for PIs selection

1.3.1 *General description*

The methodology used for the rehabilitation PIs included a sequence of steps as follows:

- selection of a subset of IWA PIs considered a priori appropriate for rehabilitation;
- preparation and distribution of a questionnaire to evaluate the relevance grade and the difficulty of assessment of each indicator and to get comments, suggestions and possible proposals for new indicators;
- processing and comprehensive meeting discussions of the questionnaire results and suggestions;
- final definition of a panel of rehabilitation PIs.

1.3.2 Selection of a subset of IWA PIs

The main criterion used has been to select indicators that can inform about issues that have influence on rehabilitation needs, effects and expenditures. The set of PIs selected from the IWA system, according to the requirements previously referred includes: 15 operational indicators dealing with sewer cleaning, rehabilitation, inflow/infiltration/exfiltration and failures; 3 physical indicators dealing with sewers surcharging, 4 environmental indicators dealing with wastewater overflows and sediments from sewers; 8 quality of service indicators dealing with flooding, interruptions of service and complaints; and 5 economical and financial indicators dealing with costs and investments.

1.3.3 *Questionnaire structure and participants*

The focus of the questionnaire was the initial set of 35 PIs selected from the IWA system of performance indicators for wastewater services (Matos *et al.*, 2003a). The main questions raised were:

- "how relevant is each indicator, based on the classification: essential, important, useful, and irrelevant";
- "how is the knowledge /perception of the assessment of the indicator: assessed/ not assessed";
- comments, suggestions and proposals for new PIs were additionally requested.

The questionnaire was sent out to the 15 Partners and the 20 End-users as illustrated in Figure 1.1. The 15 Partners are: Sintef, NTNU, Alborg University, Cemagref-Engees, WRc, Marne-la-Vallée University, Dresden University, Brno University, Budapest University, Bologna University, Ferrara University, Palermo University, Clabsa, LNEC and CSIRO. Four of them replied in association (Sintef together with NTNU, and Bologna University together with Ferrara University), resulting in completed 13 questionnaires.

The 20 End-users, answered individually, leading to 20 answers as follows: Trondheim, Bergen, Baerun, Stavanger, Oslo (in Norway), Siurliai (Lituania), Aalborg (Sweden), Severn-Trent(UK), Nantes (France), Dresden (Germany), Budapest and Budapest Vaziviz (Hungary), Brno (Check Republic), AGAC, AMAP and ACOSEA (Italy), Clabsa (Barcelona), Oeiras-Amadora (Portugal) and Newcastle and Melbourne (Australia).

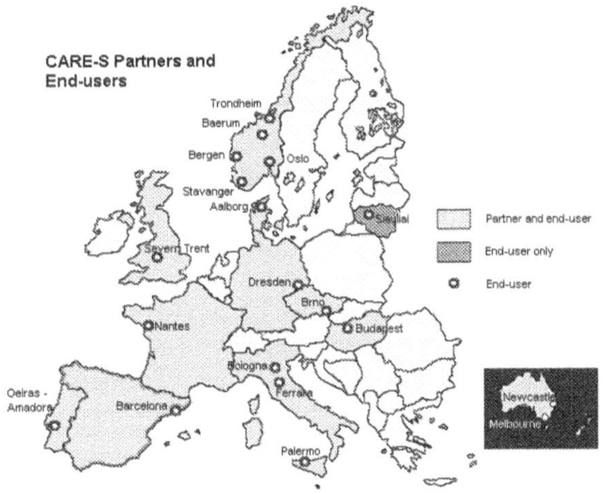

Figure 1.1 CARE-S Partners and End-users

1.3.4 Data processing

For processing of data, three different clusters – partners, end-users and the total sample – were considered in separate for each PI, to allow for a better understanding of the main differences. The data processing included the following main tasks:

- Statistical analysis of the relevance given to each PI (for the different clusters);
- Statistical analysis of the assessment/no assessment of PIs (for the different clusters);
- Relative ranking of the PIs based on Pareto histograms, for the category "relevance", according to the following group of answers: essential plus important; useful, irrelevant or not applicable;
- Relative ranking of the PIs, based on Pareto histograms, for the category "assessment" according to the following group of answers: "Assessed" and "not assessed";
- A comparative and global evaluation (including all PIs) with regard to "relevance" and "assessment", for both Partners and End-users.

The procedure has been applied to all relevant issues, namely: environmental, physical, operational, quality of service, and economical and financial.

1.4 DISCUSSION OF RESULTS

1.4.1 General remarks

The main overall conclusions drawn from the questionnaire results were as follows:

- There was a clear validation of the proposal regarding the pre-selected set of 35 rehabilitation PIs. In fact all PIs have got the classification of "essential" or "important" by at least 2 responses. Pareto histograms show that the lowest twenty-five percent of PIs that are ranked as "essential or important" are highly ranked as "useful" by the others. A fact to retain is that the lowest ranked "essential or important" PIs are, in general, considered to be easily "assessed".
- Partners and End-users are "two clusters" that appear to have different perception regarding PIs. Therefore it was clearly important to involve End-users in an early stage.
- In general Partners tend to over-estimate the importance of PIs in comparison with End-users while Partners tend to be "more optimistic" than End-Users concerning the PI assessment.

1.4.2 Analysis of the relevance given to each PI

A detailed analysis of the relevance given to each PI was carried out and the answers received for each indicator (and each group) were graphically represented as a percentage of the answers received (Matos *et al.*, 2003b). Partners, end-users and the total sample are presented in separate for each PI to allow for a better understanding of the main differences. The following considerations arise from the results:

Operational indicators: In general, the operational indicators proposed are considered by a clear majority of partners and end-users (60% to 90%) as being "essential" or "important". Exception, among the fifteen operational PIs, are three indicators related to manhole replacement, manhole covers replacement and service connections replacement (Figure 1.2).

Physical indicators: the physical indicators are considered, in general, by a majority of partners and end-users (50% to 70%) as being "essential" or "important" (Figure 1.3).

Environmental indicators: three of the environmental indicators (those dealing with surcharging of sewers) are considered by a majority of partners and end-users (50% to 70%) as being "essential" or "important". Exception is the indicator related to sediments in sewers, which are considered relevant or important by close to 25% of end-users and 30% of partners (Figure 1.4).

Quality of Service indicators: in general, the quality of service indicators dealing with the service itself (namely flooding of properties) are considered by the majority of partners and end-users (60%-70%) as being "essential" or "important". The indicators dealing with complaints (namely blockage and flooding complaints) are considered "relevant" by only 15% of partners and end-users and "important" by 25-30 % of them (Figure 1.5).

Financial indicators: financial indicators are considered in general by a clear majority of partners and end-users (50% to 90%) as being "essential" or "important" (Figure 1.6).

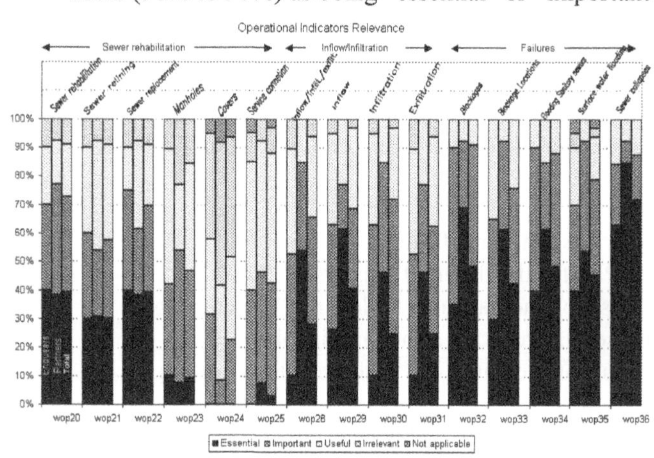

Figure 1.2 Relevance of operational indicators

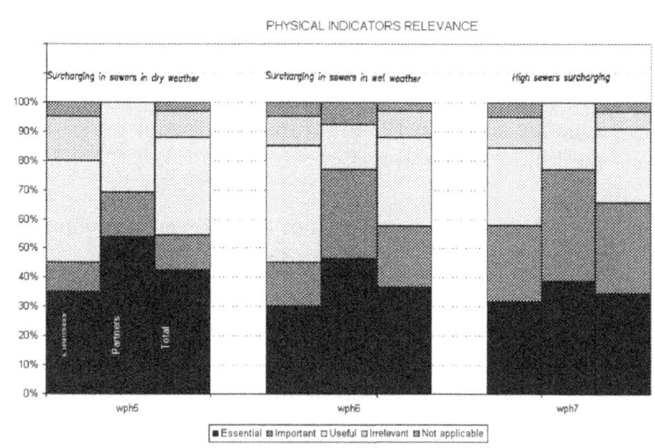

Figure 1.3 Relevance of physical indicators

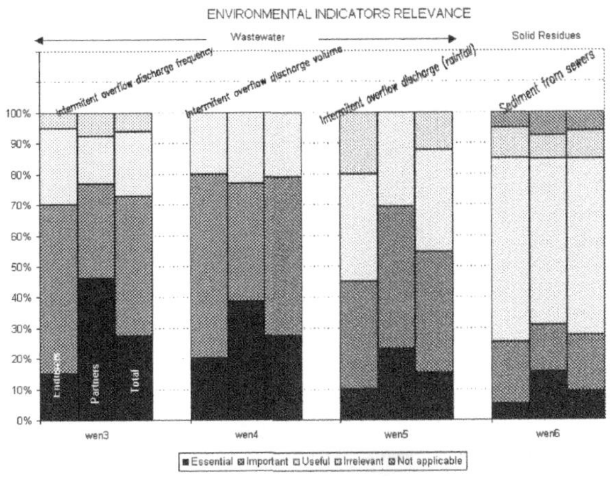

Figure 1.4 Relevance of environmental indicators

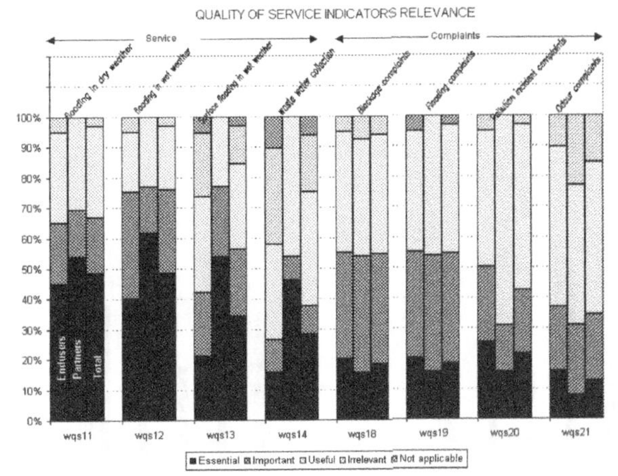

Figure 1.5 Relevance of quality of service indicators

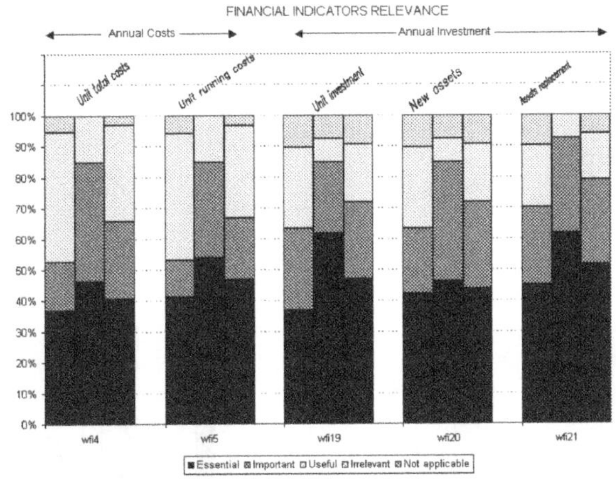

Figure 1.6 Relevance of financial indicators

1.5 ANALYSIS OF THE EASE OF ASSESSMENT GIVEN TO EACH PI

The same analysis was carried out regarding the ease of assessment of the indicators (Matos *et al.*, 2003b). The following considerations arise from the results:

Operational indicators: In general, the operational indicators are considered "assessed" by a clear majority of partners and end-users (50% to 95%). Exceptions, among the fifteen operational PIs, are the indicators related to infiltration, inflow and exfiltration were the expectancy of assessment is clearly low: between 5% and 20% of the answers (Figure 1.7).

Physical indicators: the physical indicators are considered "assessed" by 30% to 40% of the partners and end-users. The behaviour seems to be very similar between the two clusters (Figure 1.8).

Environmental indicators: the environmental indicators are considered "assessed" by 45 % to 55% of the partners and by 20% to 45% of the end-users. The expectancy shows in this case a different behaviour between partners and end-users (Figure 1.9).

Quality of Service indicators: in general, quality of service indicators is considered assessed by 45% to 85% of end-users and by 55% to 95% of partners. End-users tend to be less optimistic then partners regarding the assessment of indicators dealing with the service itself (e.g. flooding of properties) and more optimistic than partners regarding "complaints of the service" (Figure 1.10).

Financial indicators: financial indicators are considered both from partners and end-users easy to assess. Partners and end-users are in line, specifically regarding 'investment' PIs. Light differences are shown as regards to "costs" PIs. Answers vary from 90% to 100% (Figure 1.11).

In general, the operational indicators proposed are considered by a clear majority of partners and end-users (50% to 90%) as being "essential" or "important". Exception, among the fifteen operational PIs, are the three indicators related to manhole replacement, manhole covers replacement and service connections replacement. They are also considered "assessed" by a clear majority of partners and end-users (50% to 100%). Exceptions, among the fifteen operational PIs, are the indicators related to infiltration, inflow and exfiltration were the expectancy of assessment is clearly low: between 5% and 20% of the answers (Figure 1.7).

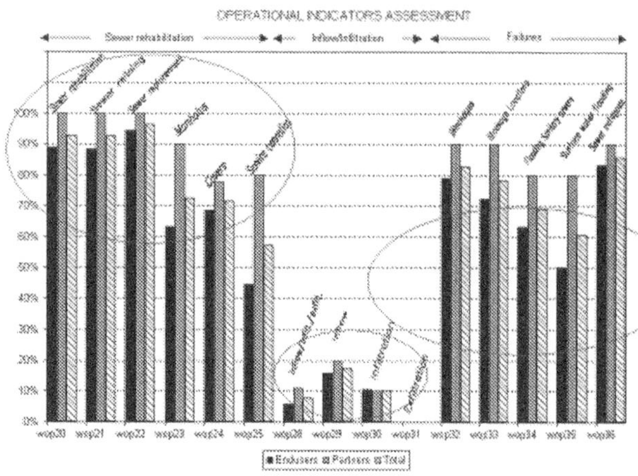

Figure 1.7 Operational indicators assessment

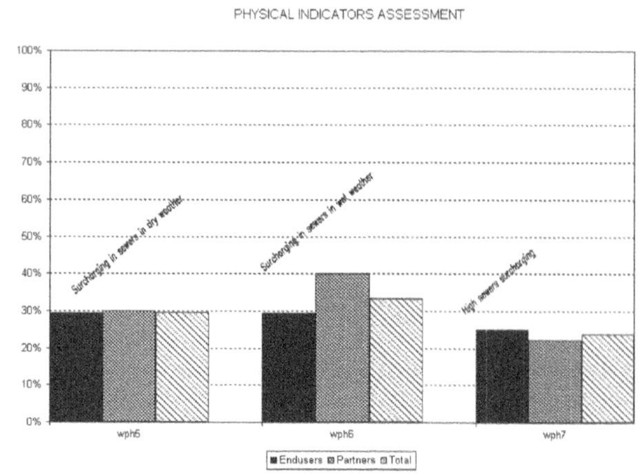

Figure 1.8 Physical indicators assessment

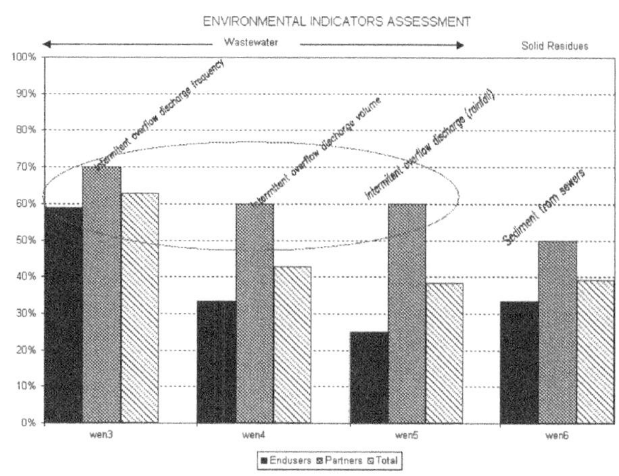

Figure 1.9 Environmental indicators assessment

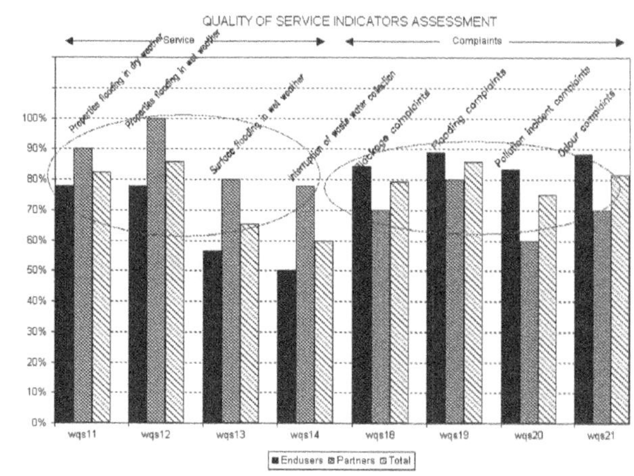

Figure 1.10 Quality of service indicators

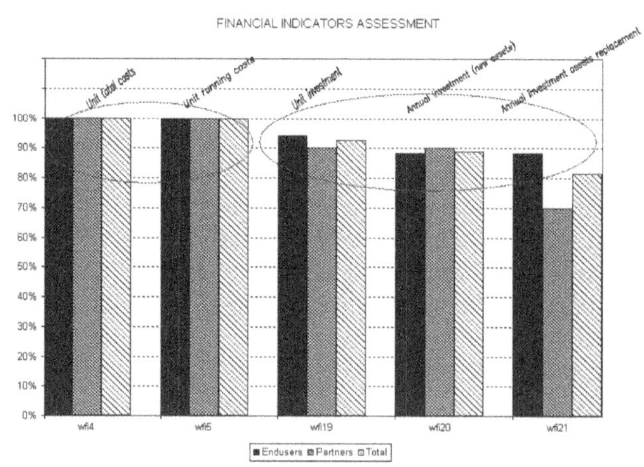

Figure 1.11 Financial indicators assessment

1.6 PI TOOL/S SOFTWARE

The PI Tool/S is a computer program developed by LNEC (Praça and Coelho, 2003). It makes available to the CARE-S sewer network rehabilitation procedure an information system for the calculation, storage, update, retrieval and analysis of PIs and their associated information. Apart from the values of indicators and their calculation data, the system provides complete information on all their features, properties and relationships, and offers graphical analysis capabilities. The PI Tool/S offers a facility to graphically represent PI and UI values, including 3 types of charting: time series graphs, dataset comparison graphs and PI comparison graphs. The PI Tool/S makes use of the common CARE-S database standard. The PI Tool/S is published either as a stand-alone program or as an integrated facility within the CARE-S software package.

1.7 PI FIELD-TESTING

1.7.1 General description

The PI field-testing involved the 20 project End-users who were invited to use the software PI Tool/S, input data required to assess performance indicators and to report on the difficulties. End-users were requested to assess as many indicators as possible, even if they did not plan to use them on a routine basis in the future. The results of the PIs field-testing are presented in Matos *et al.* (2004b).

One important aspect of the PI data collection and comparison is the assessment of their application feasibility. Figure 1.12 shows the number of indicators assessed by each participant, from a total of 41. The mean number of PI calculated per participant is 21, 50% having a number of answers in the range of 19 to 26 PI. According to the replies to the questionnaire delivered, most difficulties are related with the fact that despite data exist, it is not recorded in a way that can be easily used as input to the PI Tool/S. In other cases data required are not traditionally collected, but there are no technical difficulties in the process. The results of the PI assessment for the different end-users were examined and discussed. The outcome results from end-users are considered confidential and therefore each end-user is referred to with a code form and with the author's guaranty that data will not be disseminated in any other form.

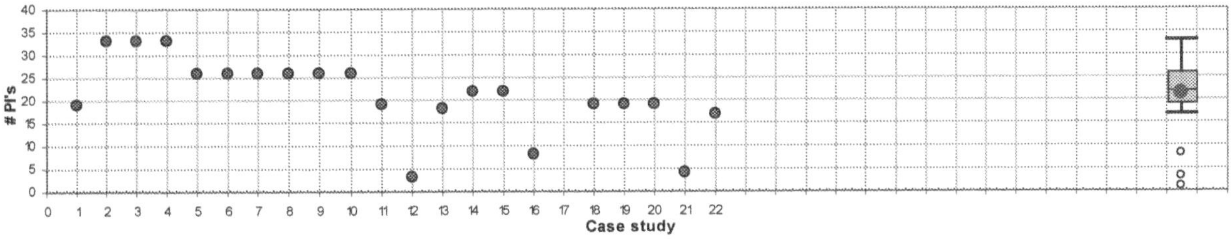

Figure 1.12 Total number of performance indicators (PI) assessed per undertaking

Data provided and the informal exchanges of information show that there are significant differences on the characteristics of the networks analysed, for instance in the size of the systems. Answers collected were not enough to allow splitting the data into different data sets so only the whole sample was analysed. The differences in size of the networks analysed, is shown in terms of the total sewer length in Figure 1.13. The mean sewer length of the case studies analysed was 1048 km. There is one case study with sewer length less than 17 km and there are two case studies with sewer length greater than 1563 km.

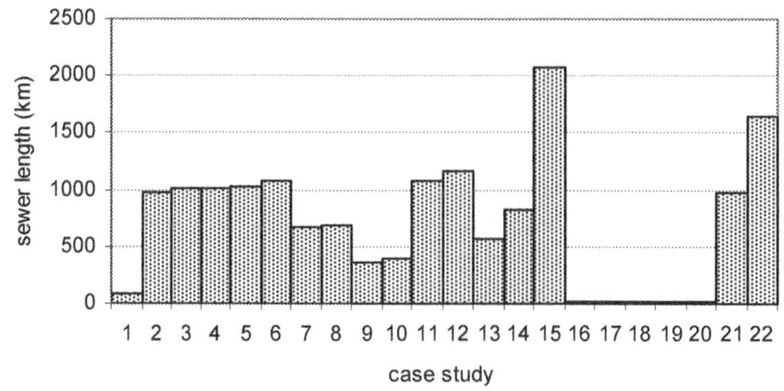

Figure 1.13 Size of the networks analysed

1.7.2 Results

The global statistics are presented in the form of box & whiskers plots, for all the indicators having at least 7 replies received. When outliers or extreme observations exist, the minimum and maximum observations correspond to the non-outlier range, but the other statistics are calculated with the entire sample, including outliers and extreme observations

A total of 445 valid results were obtained, distributed over the 41 PI (Figure 1.3). For the subgroup inflow/infiltration/exfiltration no answers were obtained at all. The indicators sQs4 (interruption of wastewater collection and transport services), sEn1 (overflow discharge frequency), sEn2 (overflow discharge volume), sEn3 (duration of overflow discharge) and SEn4 (overflow discharge related to rainfall) had less than seven results (Cardoso *et al.*, 2004).

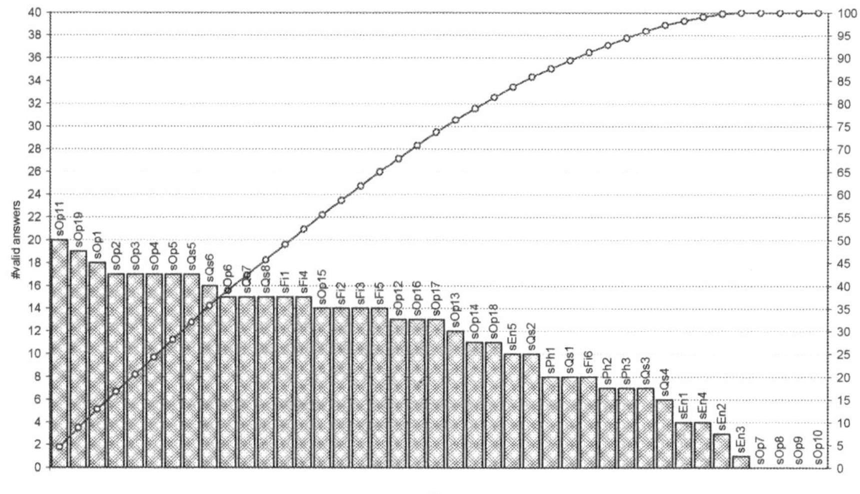

Figure 1.14 Total number of valid results per performance indicator (PI)

Table 1.1 PI global statistics (* represents an outlier)

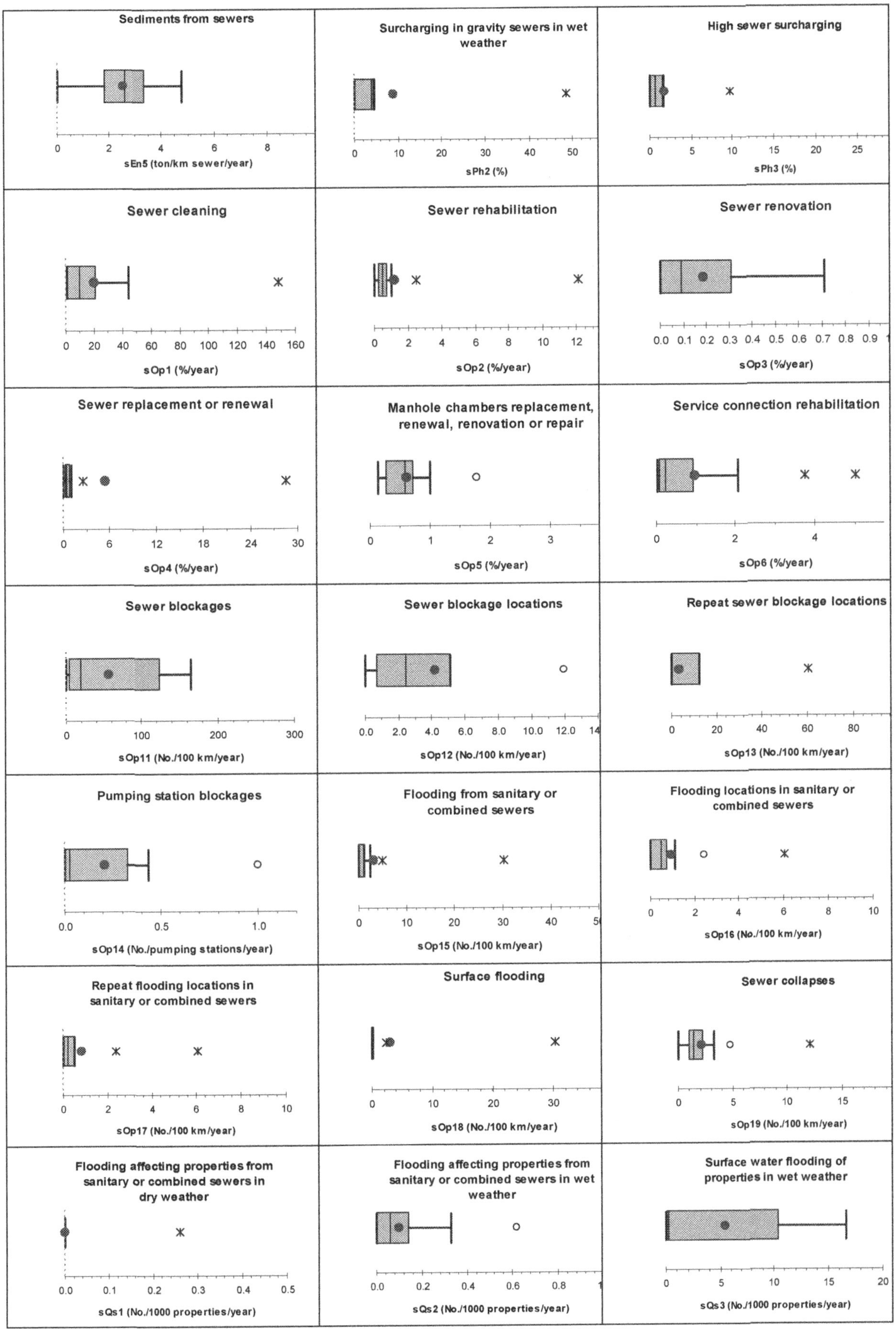

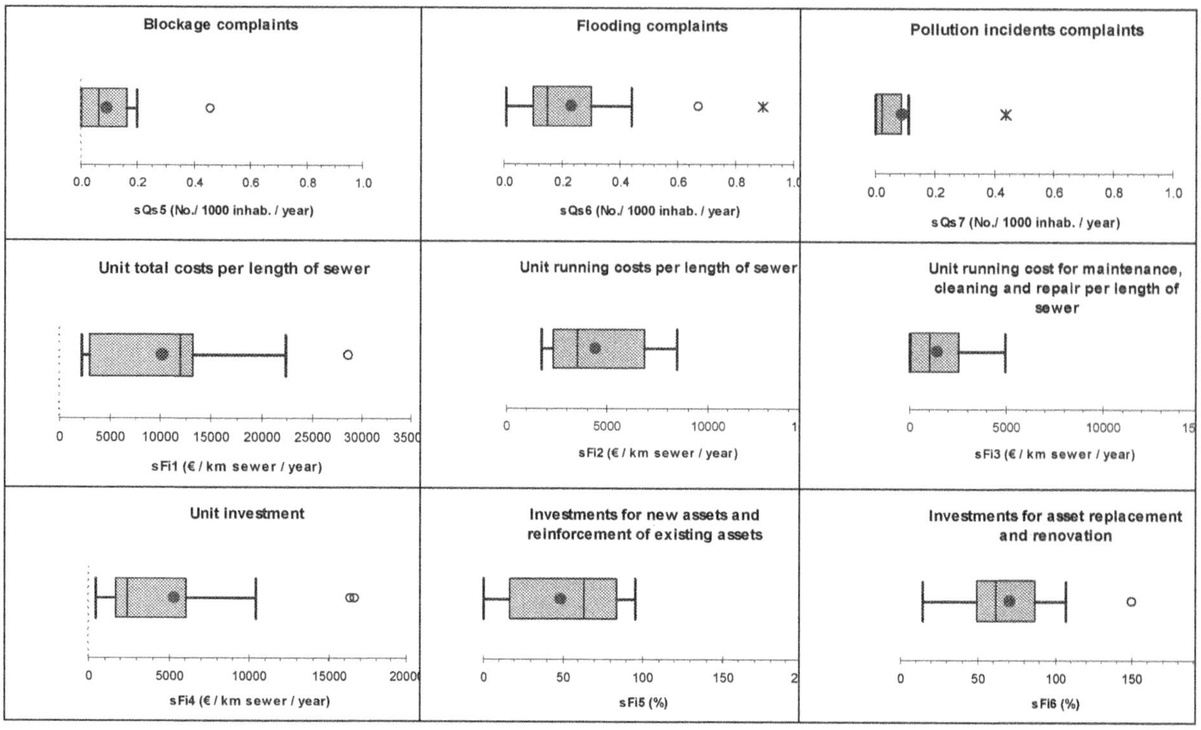

Table 1.2 Extract of PI results

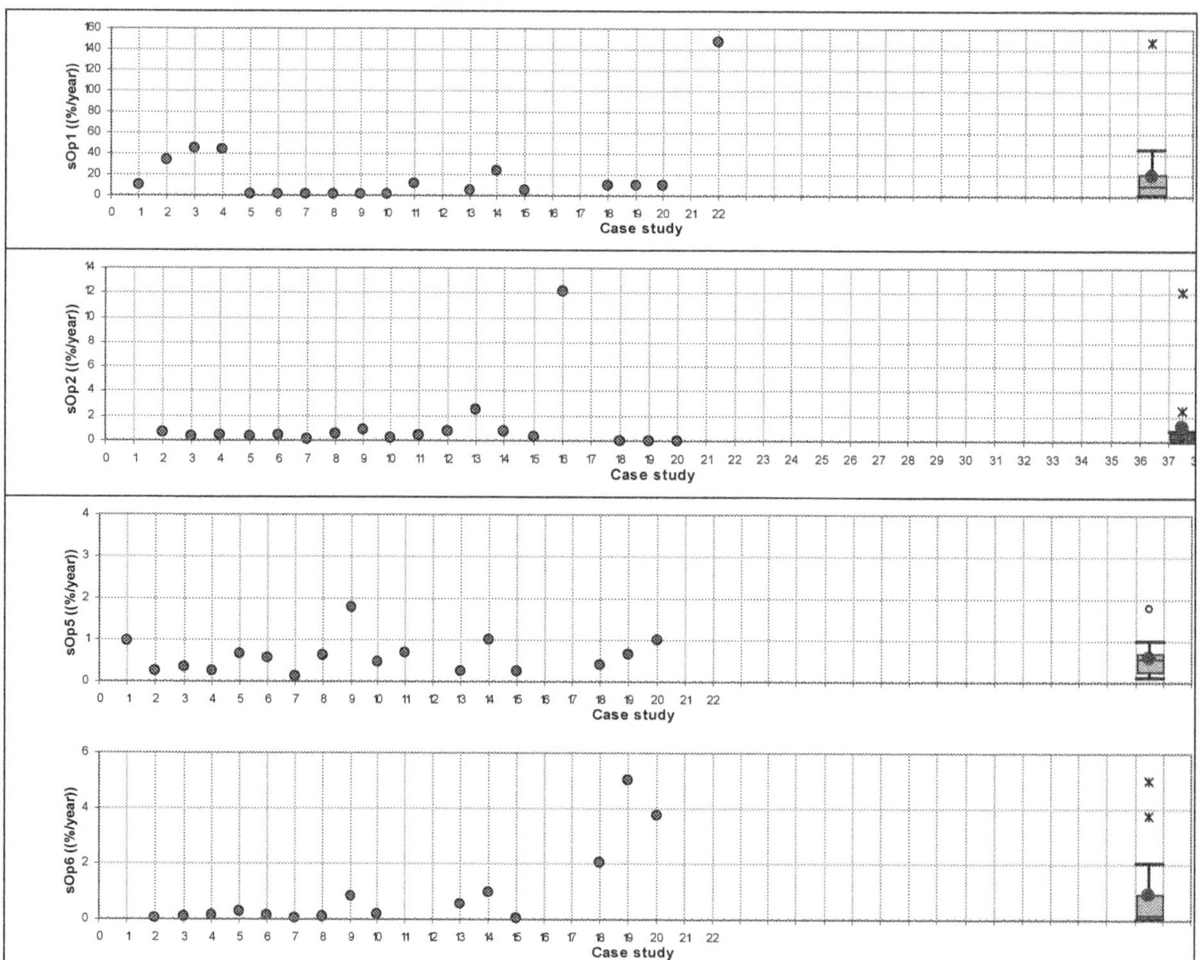

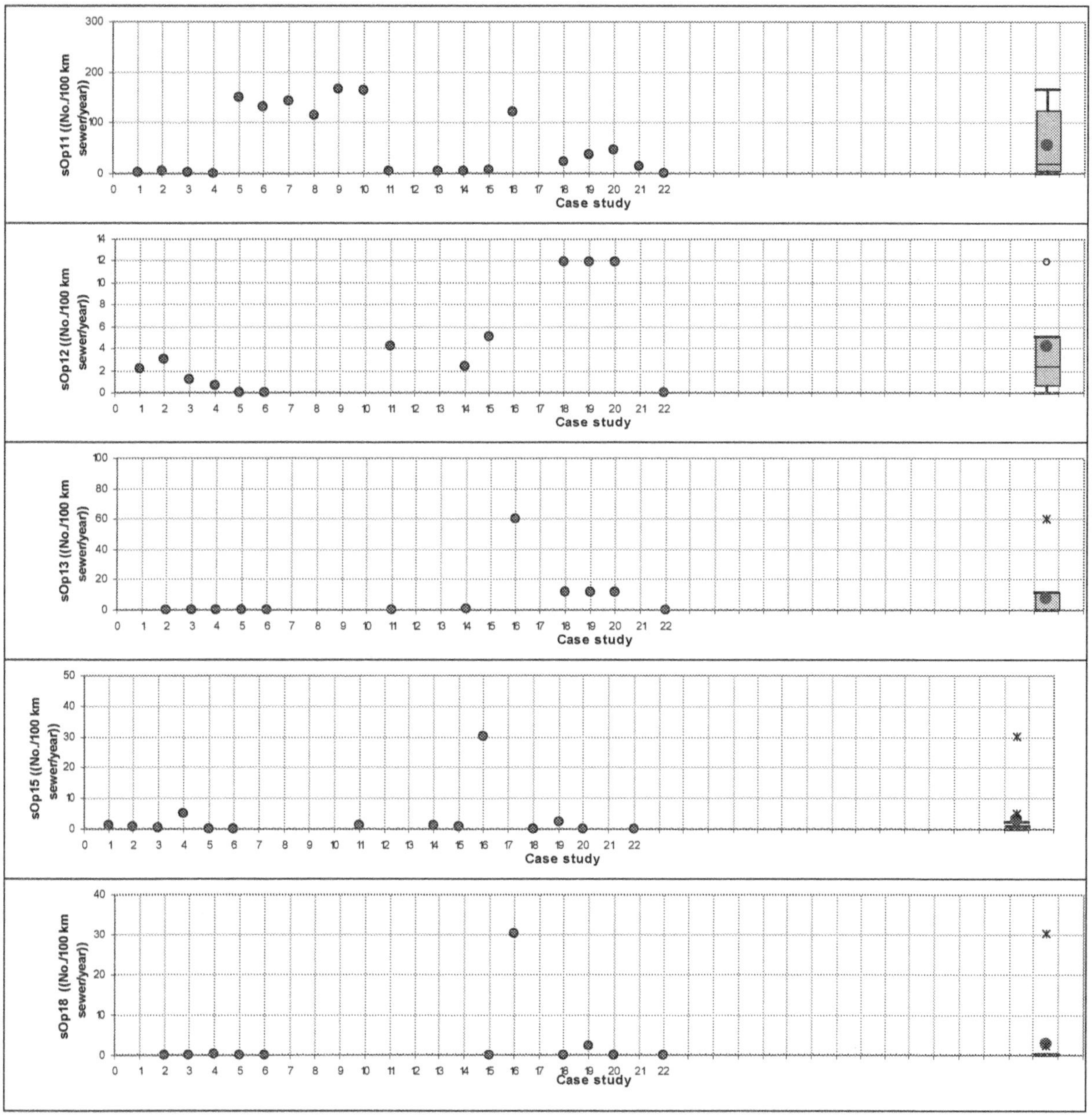

Table 1.2 PI results (cont.)

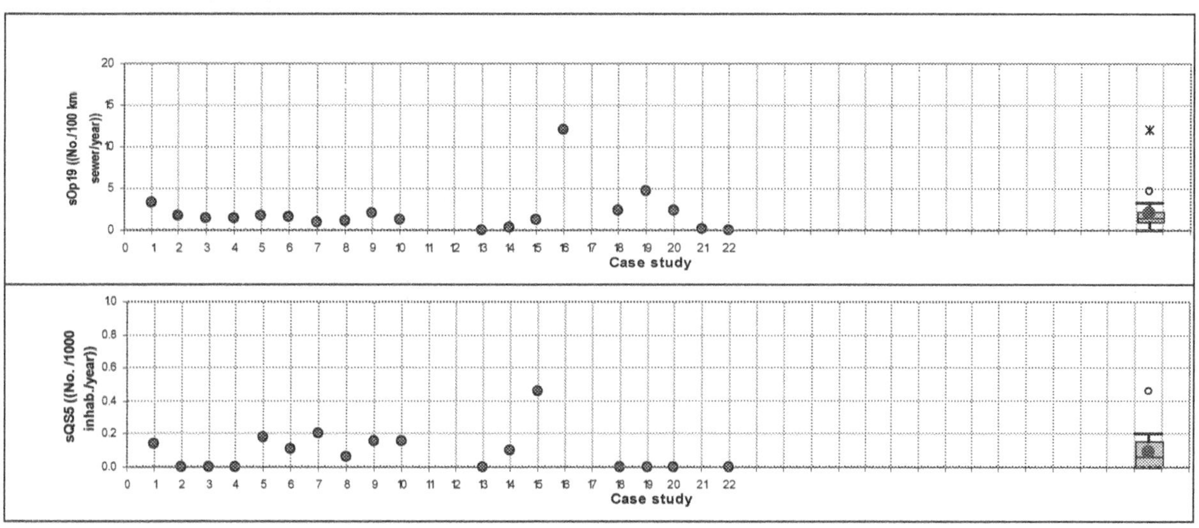

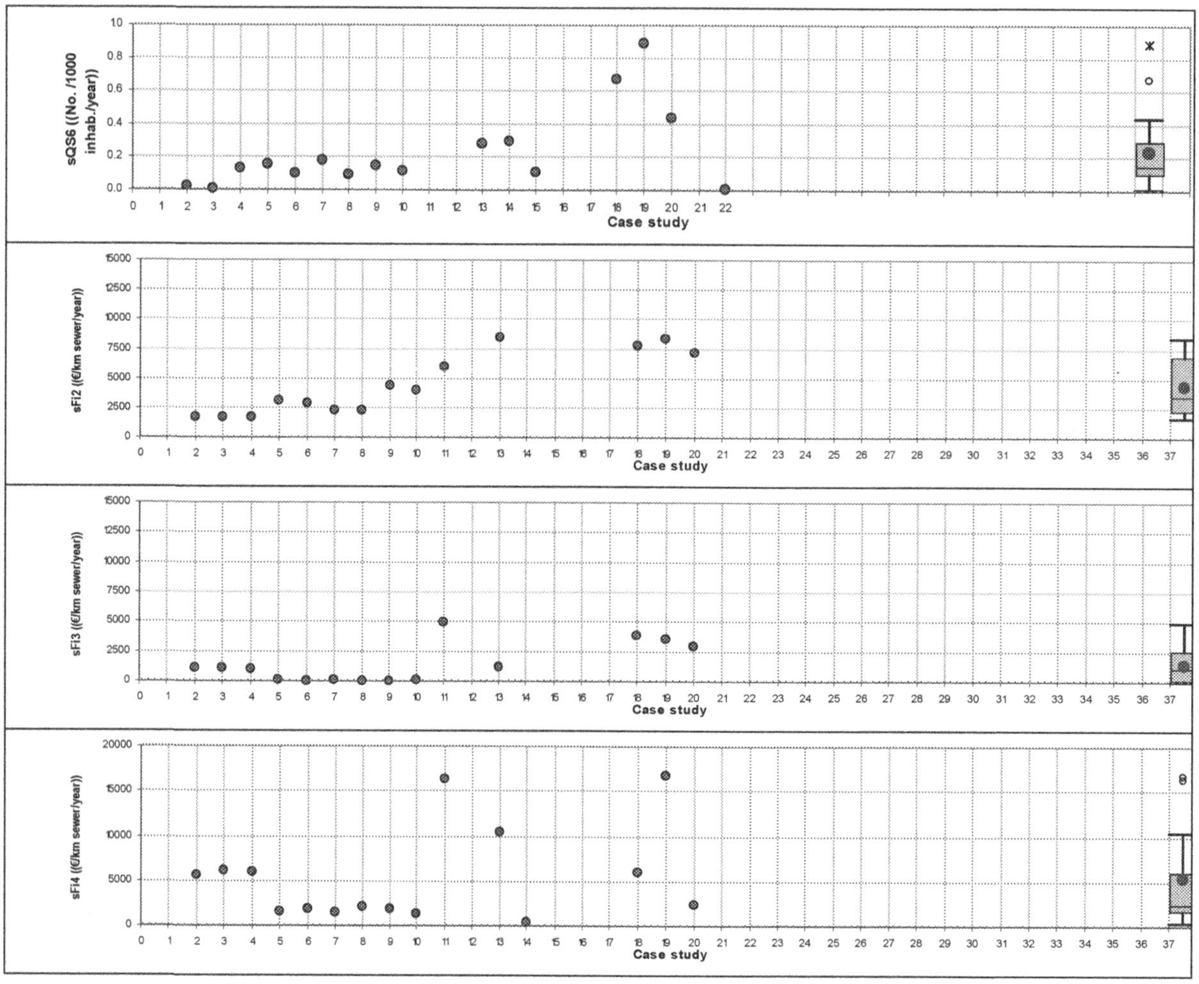

From the results analysis of the field-testing it is possible to identify a relationship between the following indicators (Cardoso *et al.*, 2004):

- sOp1 (*sewer cleaning*) with sOp11 (*sewer blockages*) and sOp12 (*sewer blockages locations*) and sOp13 (*repeat sewer blockages locations*): for higher values of sOp1, one can observe that sOp11, sOp12 and sOp13 present lower values. This means that systems with higher percentages of cleaning show less blockages per length of sewer as well as lower number of blockages locations and repeated blockages locations per length of sewer.
- sOp2 (*sewer rehabilitation*) with sOp19 (*sewer collapses*): for higher values of sOp2, one can observe that sOp19 presents lower values. This means that systems with higher percentages of rehabilitation show less number of collapses per length of sewer.
- sOp11 (*sewer blockages*) with sQs5 (*blockage complaints*): as expected for higher values of sOp11 one can observe also higher values in sQs5. This means that systems with higher sewer blockages receive more complaints related to blockages.
- sOp15 (*sewer flooding from sanitary or combined sewers*) and sOp18 (*surface flooding*) with sQs6 (*flooding complaints*): for higher values of sOp15 and/or sOp18 one can observe also higher values of sQs6. This means that systems with higher sewer flooding receive more complaints related to flooding.
- sFi2 (*unit running costs per length of sewer*) with sOp2 (*sewer rehabilitation*), sOp5 (*manhole chambers replacement, renewal, renovation and repair*) and sOp6 (*service connection rehabilitation*): there is a tendency that for increasing values of sFi2, also higher values for sOp2 and/or sOp5 and/or sOp6 are observed, and vice-versa. This means that higher percentages of sewer and service connection rehabilitation and manhole chambers replacement, renewal, renovation and repair have impacts on the running costs.
- sFi3 (*unit running costs for maintenance, cleaning and repair per length of sewer*) with sOp1 (*sewer cleaning*): for higher values of sFi3, one can observe that sOp1 presents also higher values and vice-versa. This means that higher percentages of cleaning have impacts on the running costs for maintenance.

- sFi4 (*unit total costs per length of sewer*) with sOp11 (*sewer blockages*) and sOp19 (*sewer collapses*): there is a tendency that for increasing values of sFi4, one can observe that sOp11 and sOp19 present lower values and vice-versa. This means that higher unit investments have beneficial impacts on the blockages and collapses.

1.8 FINAL PANEL OF PIS FOR REHABILITATION

The final PI listing established for the CARE-S is presented in Table 1.3. As a result of the field-testing some PI changed with regard to the first set and are in bold.

Table 1.3 – Final listing of PIs for rehabilitation

ENVIRONMENTAL INDICATORS (sEn)	
Wastewater	**IWA PI**
sEn1 - Overflow discharge frequency (No./overflow device/year)	wEn3
sEn2 - Overflow discharge volume (m^3/overflow device/year)	wEn4
sEn3 - Duration of overflow discharge ((hours)/overflow device/year)	-
sEn4 - Overflow discharge related to rainfall (%/year)	wEn5
Sediments	
sEn5 - Sediments from sewers (ton/km sewer/year)	wEn12
PHYSICAL INDICATORS (sPh)	
Sewers	
sPh1 - Surcharging in gravity sewers in dry weather (%)	wPh5
sPh2 - Surcharging in gravity sewers in wet weather (%)	wPh6
sPh3 - High sewer surcharging (%)	wPh7
OPERATIONAL INDICATORS (sOp)	
Sewer Cleaning	
sOp1 - Sewer cleaning (%/year)	wOp2
Sewer Rehabilitation	
sOp2 - Sewer rehabilitation (%/year)	wOp21
sOp3 - sewer renovation (%/year)	wOp22
sOp4 - sewer replacement or renewal (%/year)	wOp23
sOp5 - Manhole chambers replacement, renewal, renovation or repair (%/year)	wOp25
sOp6 - Service connection rehabilitation (%/year)	wOp27
Inflow / Infiltration / Exfiltration (I/I/E)	
sOp7 - Inflow / Infiltration / Exfiltration (I/I/E) (%)	wOp30
sOp8 - Inflow (m^3/km/year)	wOp31
sOp9 – Infiltration (m^3/km/year)	wOp32
sOp10 - Exfiltration(m^3/km/year)	wOp33
Failures	
sOp11 - Sewer blockages (No./100 km sewer/year)	wOp34
sOp12 - Sewer blockage locations (No./100 km sewer/year)	wOp35
sOp13 - repeat sewer blockage locations (No./100 km sewer/year)	-
sOp14 – Pumping station blockages (No./100 km sewer/year)	wOp36
sOp15a – Flooding from sanitary sewers (No./100 km sewer/year)	-
sOp15b – Flooding from combined sewers (No./100 km sewer/year)	-
sOp16a – Flooding locations in sanitary sewers (No./100 km sewer/year)	-
sOp16b – Flooding locations in combined sewers (No./100 km sewer/year)	-
sOp17a - Repeat flooding locations in sanitary sewers	-
sOp17b - Repeat flooding locations in combined sewers	-
sOp18 - Surface flooding (No./100 km sewer/year)	wOp39
sOp19 - Sewer collapses (No./100 km sewer/year)	wOp40

QUALITY OF SERVICE INDICATORS (sQS)	
Flooding	
sQS1a - Flooding affecting properties from sanitary sewers in dry weather (No./1000 properties/year)	-
sQS1b - Flooding affecting properties from combined sewers in dry weather (No./1000 properties/year)	-
sQS2a - Flooding affecting properties from sanitary sewers in wet weather (No./1000 properties/year)	-
sQS2b - Flooding affecting properties from combined sewers in wet weather (No./1000 properties/year)	**wQS11**
sQS3 - Surface water flooding of properties in wet weather (No./1000 properties/year)	wQS14
Interruptions	
sQS4 - Interruption of wastewater collection and transport services (%)	wQS15
Complaints	
sQS5 – Blockage complaints (No. /1000 inhab./year)	wQS20
sQS6 - Flooding complaints (No. /1000 inhab./year)	wQS21
sQS7 - Pollution incidents complaints (No. /1000 inhab./year)	wQS22
sQS8 – Odour complaints (No. /1000 inhab./year)	wQS23
ECONOMIC AND FINANCIAL INDICATORS (sFi)	
Costs	
(*) sFi1 - Unit total cost per length of sewer (€/km sewer/year)	wFi6
(*) sFi2 - Unit running cost per length of sewer (€/km sewer/year)	wFi8
(*) sFi3 - unit running cost for maintenance, cleaning and repair per length of sewer (€/km sewer/year)	-
Investment	
(*) sFi4 - Unit investment (€/km sewer/year)	wFi27
sFi5 - investments for new assets and reinforcement of existing assets (%)	wFi28
sFi6 - investments for asset replacement and renovation (%)	wFi29

(*) *Note that if these costs are referred to inhabitants, than this indicator shall be expressed in €/inhabitant/year*

The final Rehab Utility Information listing established for the CARE-S presented in Table 1.4, where the changed information is marked in bold:

Table 1.4 Rehab Utility Information

ENVIRONMENTAL DATA	**IWA PI**
• Sewer sediments – (ton)	wA20
• Number of overflow discharges - (No.)	wA24
• Volume of overflow discharges - (m^3)	wA25
• Duration of overflow discharges - (hours)	-
• Rainfall volume - (m^3)	wA26
PHYSICAL ASSETS DATA	
• Catchment area - (km2)	CI
• impermeable catchment area - (km2)	CI
• Total sewer length - (km)	wC1
• combined sewers length - (km)	CI
• sanitary sewers length - (km)	CI
• stormwater sewers length - (km)	CI
• pump mains length - (km)	CI
• other sewers length - (km)	CI
• Tidal sewers - (km)	CI
• Expansion of sewer network - (km/year)	CI

• Sewer materials	CI
• clay sewers - (km)	CI
• asbestos cement sewers - (km)	CI
• concrete sewers - (km)	CI
• polyvinyl chlorine sewers - (km)	CI
• polyethylene sewers - (km)	CI
• iron sewers - (km)	CI
• steel sewers - (km)	CI
• stone sewers - (km)	CI
• brick sewers - (km)	CI
• other known material sewers - (km)	CI
• unknown material sewers - (km)	CI
• Sewer diameters or equivalent	CI
• Dia ≤ 150 mm - (km)	CI
• 150 < Dia ≤450 mm - (km)	CI
• 450 < Dia ≤ 900 mm - (km)	CI
• 900 < Dia ≤ 1200 mm - (km)	CI
• 1200 < Dia ≤ 2200 mm - (km)	CI
• Dia > 2200 mm - (km)	CI
• Unknown diameter - (km)	CI
• Sewer age	CI
• sewers laid after 1995 - (km)	CI
• sewers laid between 1985 and 1994 inclusive - (km)	CI
• sewers laid between 1975 and 1984 inclusive - (km)	CI
• sewers laid between 1950 and 1974 inclusive - (km)	CI
• sewers laid between 1925 and 1949 inclusive - (km)	CI
• sewers laid before 1925 - (km)	CI
• unknown age sewers - (km)	CI
• Sewers location	
• under flexible roadway – (Code UFR)	-
• under rigid roadway – (Code URR)	-
• under sidewalk – (Code USW)	-
• under green areas – (Code UGA)	-
• Sewers installation depth – (m)	
• Sewers trench width – (m)	-
• Bedding soil type	
• categories to be defined – (alphanumeric)	-
• Backfilling soil type	
• categories to be defined – (alphanumeric)	-
• Average closeness to trees – (m)	-
• Type of joints	
• rigid joints – (Code RJ)	-
• flexible joints – (Code FJ)	-
• Surcharged sewers in dry weather – (m)	wC2
• Surcharged sewers in wet weather – (m)	wC3
• Highly surcharged sewers – (m)	wC4
• Overflow devices - (No.)	wC19
• Manhole chambers - (No.)	wC21
• Gully pots - (No.)	wC22
• Connected properties - (No.)	wC28
• Service connections - (No.)	wC29
• domestic connections - (No.)	CI
• industrial connections - (No.)	CI
•	

Wastewater Storage	
• Storage tanks - (No.)	wC23
• Total storage volume - (m^3)	CI
• Sewer system pumping stations - (No.)	wC9
Pumped wastewater - (m^3)	CI
Technological Resources Computerised information systems (IT=Information Technology)	
• Routine use of Information Technology to support	
• inspection (including visual and CCTV) – (yes/no)	CI
• maintenance – (yes/no)	CI
• customer complaints – (yes/no)	CI
• updated mapping - (km)	CI
• digital mapping - (km)	CI
OPERATIONAL DATA	
• Sewer cleaning - (km)	wD2
• Sewer rehabilitation - (km)	wD25
• Sewer renovation - (km)	wD26
• Sewer replacement - (km)	wD27
• Manhole chamber replacement, renewal, renovation or repair - (No.)	wD29
• Replacement of manhole covers - (No.)	wOp26
• Service connection replacement or renewal - (No.)	wD31
• Inflow volume - (m^3)	wD35
• Infiltration volume - (m^3)	wD36
• Exfiltration volume - (m^3)	wD37
• Sewer blockages - (No.)	wD38
• Sewer blockage locations - (No.)	wD39
• Repeat sewer blockage locations - (No.)	-
• Pumping station blockages - (No.)	wD40
• **Floodings from sanitary sewers - (No.)**	-
• **Floodings from combined sewers - (No.)**	-
• **Flooding locations from sanitary sewers - (No.)**	-
• **Flooding locations from combined sewers - (No.)**	-
• **Repeat flooding locations from sanitary sewers - (No.)**	-
• **Repeat flooding locations from combined sewers - (No.)**	-
• Surface floodings - (No.)	wD43
• Sewer collapses - (No.)	wD44
DEMOGRAPHY (AND CUSTOMER) DATA	
• Resident population – (Inhab.)	wE1
• Resident population connected to SE – (Inhab.)	wE4
• Resident population served by on-site systems – (Inhab.)	CI
• Collected sewage - (m^3)	wF1
• Industrial wastewater - (m^3)	CI
QUALITY OF SERVICE DATA	
• **Dry weather flooding of properties from sanitary sewers - (No.)**	-
• **Dry weather flooding of properties from combined sewers - (No.)**	-
• **Wet weather flooding of properties from sanitary sewers - (No.)**	-
• **Wet weather flooding of properties from combined sewers - (No.)**	**wF3**
• Wet weather surface flooding of properties - (No.)	wF6
• Wastewater service interruptions - (No.)	wF7
• Blockage complaints - (No.)	wF13
• Flooding complaints - (No.)	wF14

• Pollution complaints - (No.)	wF15
• Odour complaints - (No.)	wF16
ECONOMIC AND FINANCIAL DATA	
• Total costs – (€)	wG5
• External services costs – (€)	wG10
• Running costs – (€)	wG6
• Running cost for maintenance, cleaning and repair – (€)	-
• Investment in tangible assets – (€)	wG30
• Investments for new assets and reinforcement of existing assets – (€)	wG31
• Investments for assets replacement and renovation – (€)	wG32
TIME DATA	
• Assessment period – (days)	wH1

1.9 FINAL REMARKS

The field-testing of the PIs undertaken in the scope of the project CARE-S had a participation of about 50% of the end-users and the results are very relevant. However, the usefulness of the rehabilitation performance indicators must be further explored.

1.10 CONCLUSIONS

A control panel of performance indicators for rehabilitation was developed within the CARE-S project and tested by the end-users. The PI are structured in five groups such as environmental, physical, operational, quality of service and economic and financial. From the field-testing it was possible to conclude that:

- For rehabilitation PI assessment it is fundamental to establish procedures to assure that all relevant systems data is collected; e.g. none of the end-users could provide PI regarding infiltration and exfiltration, surcharging of sewers and overflow discharges because the lack of data, in spite of the great relevance of these PI considered by the end-users (Matos *et al.*, 2004).
- It would be worthwhile to establish technically supported guidance PI ranges in order to track the performance evolution of the systems, to know when to rehabilitate and to assess or predict the benefits of rehabilitation;
- PI must be interpreted in conjunction with Utility Information and External Information that define the context of the system functioning also essential if a benchmarking approach is to be applied;
- PI can be an effective tool to support rehabilitation of sewer systems.

All the contributions received were important and complementary in the definition of the final panel of PIs for rehabilitation concerning the operational, environmental, physical, quality of service and economic financial aspects.

Acknowledgments

The authors acknowledge all the undertakings that participated in this study, as well as the corresponding CARE-S partners who assisted them.

REFERENCES

Matos, M.R.; Cardoso, M.A.; Ashley, R; Duarte, P.; Molinari, A; Shulz, A. (2003a) - Performance Indicators for Wastewaters Services. IWA Publishing, London, October, ISBN: 1900222906.

Matos, R.; Cardoso, A.; Pinheiro, I.; Almeida, M.C. (2004a) - Definition and Pilot Testing of a Control Panel of Performance Indicators for Sewer Systems Rehabilitation, NOVATECH 2004, June, Lyon, France.

Cardoso, A.; Matos, R.; Pinheiro, I.; Almeida, M.C. (2004) – Performance Indicators for Sewer Networks Rehabilitation: the users perception of their relevance and ease of assessment, Paper presented to the 4th IWA Congress, Marrakech, Morocco, September 2004 (8pp).

Matos, R., Cardoso, M.A., Pinheiro, I., Almeida, M.C. (2003b) – WP1 report - Deliverable D1 - Construction of a control panel of performance indicators for rehabilitation, selection of a listing of rehabilitation PIs. Project CARE-S -

Computer Aided REhabilitation of Sewer Networks. Decision Support Tools for Sustainable Sewer Network Management, 5th Framework Programme of the European Union, LNEC, July 2003 (172 pp).

Praça, P.; Coelho, S.T. (2003) – PI Tool/S – Version 1.0.0. - User's Manual. Laboratório Nacional de Engenharia Civil. Lisboa. Portugal. November 2003 (60 pp).

Baptista, J.M., Alegre, H. (2002) - WP1 - Construction of a control panel of performance indicators for rehabilitation. Specification for the collection of the rehab PI values required for the definition of PI thresholds. CARE-W report.

Matos, R.; Cardoso, M.A.; Pinheiro, I.; Almeida, M.C. (2004) – Construction of a Control Panel of Performance Indicators – Specification for the collection of the rehab PI values and for the tentative definition of PI thresholds. CARE S-Computed Aided Rehabilitation of Sewer Netwoks, EU project EVK1-CT-2002-00106, January 2004.

Matos, R.; Cardoso, M.A.; Pinheiro, I.; Almeida, M.C. (2004) – Construction of a Control Panel of Performance Indicators – Field testing and impacts on the final listing of rehab PIs. CARE S- Computed Aided Rehabilitation of Sewer Netwoks, EU project EVK1-CT-2002-00106, October 2004.

2

Description and validation of structural condition

Dr Jes Vollertsen, Aalborg University

2.1 INTRODUCTION

All wastewater networks deteriorate over time, and this has extremely large economic consequences. This deterioration increases the probability of failures such as obstruction and collapse, which may lead to flooding, street and building damage, traffic disturbances, environmental impacts and inconveniences for users connected to this sewer. Partially collapsed sewers have reduced flow capacity, which may lead to flooding. In the worst case of collapse, large holes may result in loss of lives.

The structural condition of a pipe depends on material, construction practices, external load and wastewater characteristics. Previous research has shown that pipes lain within certain time periods have been structurally under-designed, and that construction practices within some time periods have not been appropriate, thus leading to particularly frequent failures.

The formation of hydrogen sulphide that occurs in some wastewater networks will lead to a microbiological deterioration and loss of strength for concrete sewers, and is a very important reason for rehabilitation actions. Other reasons are root intrusion and substantial in- and ex-filtration through fissures and leaky joints.

Structural analysis of sewers is normally based on results from CCTV inspection. In Europe there are several systems currently being used for the classification of individual data from such investigations. A standard European code for the description of sewer damages has been developed and recommended to the Member States. In general, the condition of sewers is classified according to its most severe damage or overall condition to determine the urgency of rehabilitation and calculate the cost of rehabilitation.

The Workpackage 2 of CARE-S supplies tools for assessing the current state of a sewer and tools for predicting its future state. Some tools allow a forecast of the year in which a sewer enters a critical class of condition, thus determining the next inspection date or rehabilitation measure.

The Workpackage 2 of CARE-S is constituted of the following tools:

- CCTVCON: a CCTV conversion program that allows conversion of data from national coding systems into the CEN coding system. The CEN coding system is chosen as the common coding system for tools relying on CCTV data.
- GompitZ: defines the relationship between the current state and the expected service time of sewer systems using CCTV classification input.
- Infiltration/Exfiltration model: allows the simulation of exfiltration of wastewater to the surrounding soil as well as infiltration of groundwater when the pipeline is below the groundwater level.

- Blockage model: helps the user to assess the probability of sewer blockages and root penetration.
- Load model: calculates stress values in rigid non-reinforced concrete pipes based on external loads and pipe characteristics.
- ExtCorr: is a model that calculates external corrosion of concrete pipes.
- WATS internal corrosion model: simulates hydrogen sulfide induced microbial corrosion of concrete sewers by means of a conceptual approach. All relevant processes in the sewer are addressed.
- Z internal corrosion model: estimates hydrogen sulfide induced microbial corrosion of concrete sewers by a risk assessment approach.

2.2 CCTVCON

General description

Structural analysis of sewers is normally based on results from CCTV inspection. In Europe there are several systems currently being used for the classification of individual data from such investigations. A standard European code for the description of sewer damages has been developed and recommended to the member States. In general, the condition of sewers is classified according to its most severe damage or overall condition to determine the urgency of rehabilitation and calculate the cost of rehabilitation.

CCTVCON concerns the visual inspection (e.g. CCTV, man entry, mirrors, photographic camera) classification systems, of which a number of types are in use.

Major systems/standards in use in Europe as well as worldwide are identified. Each system is shortly described and a general comparison made; e.g. in terms of a matrix, depicting the features contained (e.g. types of pipes and pipe materials, manholes and other structures, private connectors, rehabilitated pipes and manholes, and etceteras) and what basic approaches are applied. The extent of use is indicated. I.e. if a method is used nationwide then the country in question is specified, if it is a more broadly used model, the broader geographic use is indicated. The status of the method is indicated, i.e. if it is a national/international standard, a regionally widely accepted method, or local, somewhat accepted method.

Six major systems are identified based on use and applicability within Europe. These systems are – together with the EU system/standard (CEN standard) – included in the CARE-S prototype. The systems are thoroughly analyzed and compared to the EU standard. The comparison includes what types of damages are addressed, weighting of damages, other information regarding the sewer itself and the validity of the model to meet the needs of tools and models for predicting the probability of asset failure.

Selecting the classification system for the project

From the comparison of different classification systems and from other experiences of the project partners the EN 13508 standard was chosen as a common system for the future tasks in the project.

Application of a general standard has several advantages and disadvantages. The main advantage of using one common standard is in the field of application software development. The application programs can use data from any sources, which keeps the rules of that standard.

The main problems emerging from using a general standard are as follows: There exists a large amount of CCTV inspection data stored by different coding systems. To be able to make use of this data, translation methods for data from such systems to the general standard is needed.

There is a problem in the field of recording new sewer defects. This is only partly important in the point of view of the project, because the models included in the project are using mainly already recorded data. But a solution should be given for the future recording too.

The application of the EN13508 standard gives solution to both problems.

The different national demands on using the EN13508 standards are satisfied via the own rules of the standard. These are in the EN13508 chapter 5.4 National Equivalent Systems and in ANNEX A National Equivalent Coding System. These rules define precisely how the national users of the CEN system can keep their traditional, familiar, national codes.

Scope of application

Converting files from national coding systems into the CEN coding system.

Data requirements:

The program recognizes the different input file formats. The current version (2.0.0.1) recognizes the German, UK, Australian, Danish, Norwegian and the CEN formats.

The translation rules (tables) are in an Access database.

Modification of the program:
Every translation rule is in Access database. The database includes the national codes and CEN codes, their position in the files and some other tables like list of materials. So the most of the modifications can be done without programming, only by changing the contents of the tables in the in the database. For example adding new codes or changing the existing ones can be carried out by Access database editing without changing the program codes. Only in case of new complex rules or a complete new national code system should the program be rewritten.

2.3 GOMPITZ

This GompitZ models defines the relationship between the current state and the expected service time of sewer systems using CCTV classification input. The model is based on empirical and statistical analysis of asset data as well as physical functions to define future conditions states, taking state variables into account.

The GompitZ tool is based on a Non Homogeneous Markov Chain (NHMC) statistical approach, and its use to model the structural degradation process of sewer pipes. A NHMC parameterization is made by the Gompit analysis of CCTV inspection data. For this purpose, the degradation process is formalized as successive transitions between condition classes defined in compliance with the recently adopted European coding system EN13508-2. Gompit Analysis is a Probit Analysis like statistical method that uses the Gompitz distribution and allows taking covariates into account.

General description

Probabilistic modeling of infrastructure degradation using NHMC has been widely used in the last ten years in North America, especially for predicting the future condition of road pavement and bridges. Concerning the degradation modeling of flexible road pavement, (Curtis and Molnar, 1997) models the degradation process in the simple Markov Chain framework, whereas (Li et al., 1997) uses explicitly NHMC, with a Bayesian update technique to compute the transition probabilities. It is also worthwhile mentioning that lifetime data analysis using Weibull distribution has been proposed in France for pavement degradation modeling (Brillet, 1995), but outside the NHMC framework.

In the case of periodically inspected infrastructure like bridges, (Madanat et al., 1995) and (DeStefano and Grivas, 1998) propose a NHMC model based on the analysis of condition state increment; the model is slightly more complex than the one presented here, and perhaps more realistic, but intractable in the case of very sparsely inspected infrastructure like sewer pipelines. A simpler model has been set up in the case of sewer pipelines, with only two condition states considered, namely good and bad (i.e. needing to be rehabilitated), the transition probability being computed by assuming that the residence time in the good state is Weibull distributed (Mailhot et al., 2000). All the methods cited in this section have been initially designed to enable the comparison of different rehabilitation scenarios.

The GompitZ statistical tool is based on a Generalized Linear Regression Model, derived from the classical Probit Regression Model using the Gompertz probability distribution (alternative to the Gaussian). The resulting model can be defined as a set of equations that enables the condition states probabilities to be calculated for any age of the pipeline, conditionally on the covariates values, and possibly on the state observed at a given age (previous inspection). The model parameters are:

- "alpha" parameters, as numerous as the number of condition classes – 1;
- "beta" parameters, one per covariate with status = 1 or 2, two per covariate with status = 3 (and obviously zero per covariate with status = 0); there is always at least one beta value estimated for the time variable.

The method used to calibrate the model is the maximization of the log-likelihood of the observations (i.e. the natural logarithm of the joined probability of the condition states of the inspected pipelines). The maximization is worked out using the Levenberg-Marquardt algorithm; the initial parameter vector is estimated by the Kaplan-Meyer method.

The role of the covariates consists in modulating the survival functions in each condition state, whereas the stratum label is used to define categories of pipelines, for each of which a separate calibration (i.e. parameters estimation) is carried out. If for instance the user chooses to consider the sole variables material and diameter, and to use their crossed levels to define strata, the resulting model has no covariates and the model predictions will be at the group level. A pure at group level analysis may also be obtained by considering no stratification and the sole variables material and diameter as covariates. If a stratification e.g. by material is chosen, and diameter, length and traffic are selected as covariates, the effect of the covariates are allowed to vary across materials; the predictions are in this case at the pipeline level, in the sense that it is unlikely that 2 different pipelines could have the same values for the stratification variable and the covariates.

Multi-State Deterioration Process: the Non Homogeneous Markov Chain approach

After having defined the pipeline as statistical analysis unit, an example of definition of the successive structural deterioration states is proposed in terms of the EN13508-2 coding system. The jumps between these states can then be viewed as a multi-state deterioration process, and formalized within the discrete time NHMC framework.

Defining The Pipeline as Analysis Unit

The CCTV-Inspection unit is a sewer pipeline comprised between two manholes that will be called pipeline in the sequel. A given pipeline can be characterized by a set of endogenous variables, i.e. material, diameter, segment length and mean depth, pipe unit length, type of effluent, etc., and environmental variables, i.e. roadway traffic, type of embedding soil, location under roadway or sidewalk, etc. These variables are considered as potential explanatory variables, accounting for the pipeline initial condition state and deterioration speed, and respectively gathered into the covariates vectors Z_0 and Z_1. Vectors Z_0 and Z_1 may also include transformed covariates, or combinations (linear or not) of covariates, possibly provided by a deterministic degradation model. Some covariates may belong to both vectors.

Defining the Structural Condition States

At any time of the service life of the pipeline, its structural condition can be classified into one of a finite number m of states, 0, 1, 2, . . . , m-1, ordered to reflect the relative degrees of deterioration of the pipeline. As an illustration one could e.g. consider the following four deterioration states:

- 0 = perfectly good,
- 1 = presence of at most surface damage and/or cracks,
- 2 = presence of fractures,
- 3 = collapse.

A possible definition for the deteriorated states 1, 2 and 3 according to the EN13508-2 coding system could be as specified in table 2.1.

Table 2.1 An example of definition of Condition Classes using EN13508-2 Defect coding

Condition Classes	Defect Codes
1	BABA (Surface cracks) or BABB (Thin fissure) or BAE (Missing mortar *) or BAF (Surface damage, BAFI excluded) or BAK (Defective lining)
2	BABC (Fracture with wall pieces still in place) or BACA (Break with wall pieces still in place) or BADA (Masonry units displaced *) or BAFI (Surface damage with missing wall)
3	BAAA (Vertical deformation **) or BACB (Break with displacement) or BACC (Complete loss of structural integrity) or BADB (Missing masonry units *) or BADC (Dropped invert *) or BADD (Complete loss of structural integrity *)

* masonry pipelines ** flexible materials excluded

A slightly more complicated condition state definition, also based upon the EN13508-2 codes and compliant with the 5 classes system of the Sewerage Rehabilitation Manual (WRc, 1994), is proposed within the Care-S framework, by the Water Research Center (WRc) based in Swindon, UK. Another different classification system is in use in German municipalities, and considers 6 condition classes. Moreover a continuous observation code enables to record the pipeline length concerned by a longitudinal defect that extends over more than 1 meter. It is then feasible to attribute a default length extension of 1 meter to any punctual defect, e.g. a circumferential fissure or break. The EN13508-2 coding system enables to split the total length L of the pipeline into partial lengths. For reasons of usage simplicity, neither the WRc nor the German systems propose this feature. The growing use of automated data processes by urban utilities managers will nevertheless make this refinement more and more tractable in the near future.

Defining the Stochastic Process

The deterioration process can be modeled as the stochastic process, i.e. the time elapsed since the installation of the pipeline. The deterioration process is considered as being driven by Gompitz survival probabilities. Figure 2.1 exemplifies Gompitz survival curves, which enable to calculate the condition state probability as the difference in the values of two successive curves.

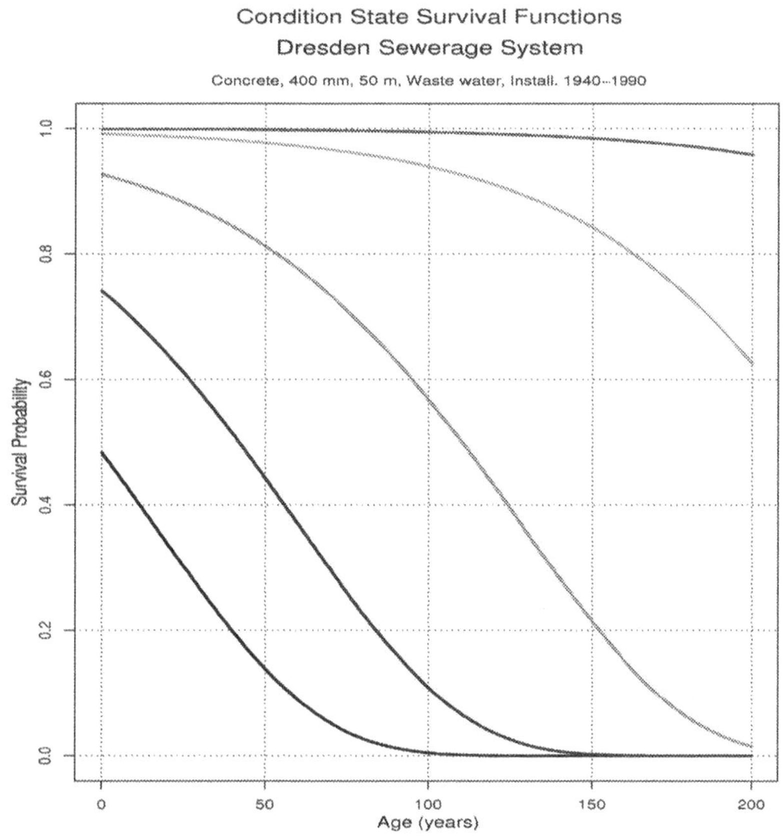

Figure 2.1 Example of Gompitz survival functions related to the 6 successive deterioration states as considered in the German standard – data from TU Dresden

Forecasting the future condition of a pipeline

The method enables the forecasting of the state probabilities vector at any time in the future, provided the current state probabilities vector was assessed, at time *t*, by CCTV inspection. If the forecasting computation concerns a pipeline that has not yet been inspected, the initial state probabilities vector may be estimated by GompitZ.

2.4 INFILTRATION/EXFILTRATION MODEL

Background

Defects in the wastewater sewer fabric (at joints and on the pipe wall) allow exfiltration of wastewater to the surrounding soil as well as infiltration of groundwater when the pipeline is below the groundwater level. Both these consequences are undesirable as they hinder the overall efficiency of wastewater systems by contributing to pollution of groundwater and reducing the efficiency of treatment plants. The primary requirement for the management problem is a method of assessing the extent of infiltration/exfiltration. The Infiltration-Exfiltration Tool (IE Tool) is designed to serve this purpose. The IE Tool is a computer program that makes available to the

CARE-S network rehabilitation procedure a system for the calculation of infiltration and exfiltration from wastewater gravity flow pipelines.

General description

The model is based on leakage through defects on different sections of the pipe wall and joints. Since the depth of flow in a sewer varies during the day, the exfiltration volume will be governed by the depth of wastewater and the position of the defects (2.2).

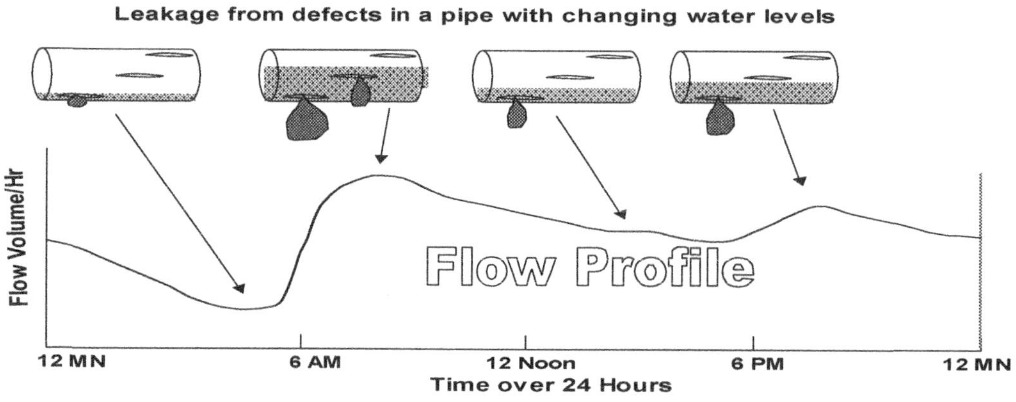

Figure 2.2 Changing levels of exfiltration with diurnal flow variation

The following assumptions were made for the development of the model:

a) The wastewater flow within gravity sewer pipes follows a daily pattern described by a hydraulic model. The 24-hour flow is considered as the sum of hourly flows during the day.

b) The wastewater flowing through pipes exfiltrates into the surrounding soil via cracks and joint defects in the pipe when the groundwater level is below the pipe invert

c) The exfiltrating wastewater forms a colmation layer (at the pipe wall and in the soil immediately in contact with the pipe). The conductivity of the colmation layer controls the rate of exfiltration.

d) When the pipe is below the groundwater level, the process is reversed and infiltration occurs through cracks and joint defects. The conductivity of the soil controls the rate of infiltration.

e) The wastewater pipe system is assumed to comprise of vitreous-clay, concrete, PVC and PE pipes. Any other pipe materials within a system are assigned to the closest within this group. The pipes are assumed to be circular in cross section.

The IE Tool makes use of the pipe asset information in the common CARE-S database. The hydraulic model in the CARE-S manager provides the wastewater flow levels in the pipes at hourly intervals. Defects in pipes are based a on a generic defects table for vitreous-clay, concrete, PVC and PE pipes.

The IE Tool is published as an integrated facility within the CARE-S software package.

Scope of application

The extent of exfiltration of wastewater from the pipes and infiltration of groundwater into the pipes are viewed as performance indicators in the CARE-S project. The program estimates exfiltration and infiltration volumes for 24-hour periods. While the flow levels in the pipes can be adjusted for different weather patterns through the hydraulic model, the groundwater level is assumed to be at a steady state, and can be only be altered by changes to the asset information data. Rehabilitation and decisions can be made based on variations to levels of infiltration and exfiltration by simulating pipe renewal or replacement through changes to the asset information file.

The basis of the exfiltration and infiltration calculations are outlined in the following sections:

Exfiltration

Exfiltration occurs when the defect in the pipe fabric (pipe wall or joint) is below the wastewater level inside the pipe. Whilst the height of the water above the defect determines the flow rate through the defect, the controlling factor is assumed to be the net thickness and conductivity of the colmation layer either side of the defect.

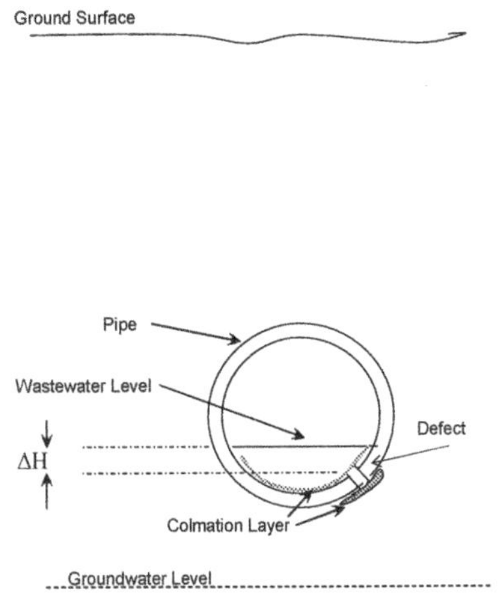

Figure 2.3 Exfiltration from sewer, principle sketch

The IE Tool takes into account;

- The changing wastewater level over a diurnal period
- The wastewater pressure at the level of the defect
- Thickness of the colmation layer
- Conductivity of the colmation layer

The leakage at any given time is:

$$Q_{(\text{Exfiltration})} = A_{\text{leak}} \cdot \Delta H \cdot (k_c / \Delta L) \ (m^3/s)$$

Where;

A_{leak} = Defect area (m^2) (from the generic defects file)
ΔH = Height of water within pipe from defect (m) (calculated by the IE tool)
k_c = Conductivity of the colmation layer (m/s) (input parameter)
ΔL = Thickness of the colmation layer (m) (input parameter)

Infiltration

Infiltration occurs when the pipe is below groundwater level. Whilst the height of the groundwater above the pipe determines the flow rate through the defect, the controlling factor is assumed to be the conductivity of the band of soil directly outside the pipe defect.

For the infiltration model, the defects are assumed to be at the top of the pipe, and thickness of the controlling band of soil to be 100 mm.

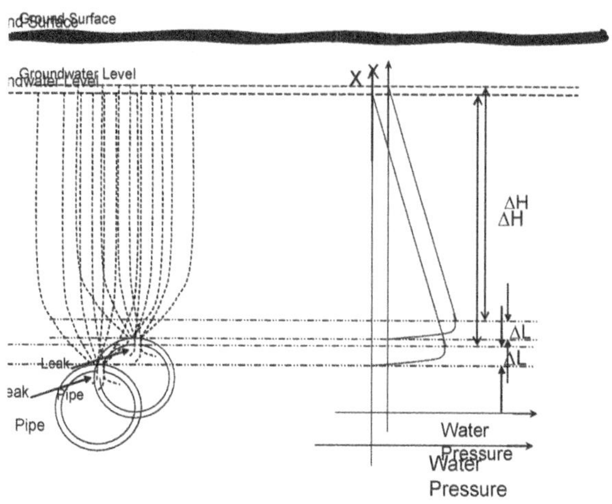

Figure 2.4 Infiltration into sewers, principle sketch

Infiltration at any time is given by:

$$Q_{(infiltration)} = A_{leak}.\ \Delta H_{GW}\ .\ (k_f / \Delta L\ ) = 10 * A_{leak}..\ \Delta H\ .\ k_f$$

Where,

Aleak.	=	Defect area (m2) (from the generic defects file)
ΔH	=	Depth of pipe below groundwater level (m) (from asset database)
ΔL	=	Thickness of limiting soil layer (assumed to be 0.1 m)
kf	=	Conductivity of soil (m/s) (from literature, input parameter))

Strengths

The asset data required by the IE Tool is very basic and is readily available from any network asset database. The depth of assets below groundwater level may not be readily available. However with a GIS incorporating groundwater levels, the relevant information could be extracted.

This gives the flexibility of applying the tool to almost any sewer network that has the basic information in a database.

Simulations for renewals and upgrades can be implemented through changes to material types or ages in the database. For instance, replacement of concrete pipes of age greater than 54 years with HDPE pipe can be simulated by changing the material codes and installed year in the database.

Limitations

The generic defects file is based on limited CCTV data on sewer pipes in Melbourne, Australia and expert opinion (the degree to which generic defects represents states of deterioration in different countries needs to be tested and validated).

2.5 THE BLOCKAGE MODEL

General description

The Blockage Tool (BT) helps the user to assess the probability of sewer blockages and root penetration. The tool is a factorial-based model where the factors are identified by statistical analysis of historical blockage data, calibrated for each user, and is to be run on all pipes in the network as an automatic routine. Typical data available for pipes, and considered of relevance for blockage modeling, are:

- Pipe ID (for pipe identification)
- Material
- Wall thickness
- Diameter
- Length
- Construction year

- Pipe type (sewer, storm water or combined)
- Joint type
- Slope and saggings
- Pipe position in the network
- Soil conditions
- Proximity to trees (for root penetration)

The tool analyses a set of blockage data and network asset data, and calculates the factors used to determine blockage probability.

Scope of application

The blockage tool calculates a blockage probability factor for each individual pipe in the network.

These results can be used for several purposes, the three main ones being:

(1) Rehabilitation planning, as input to the multi-criteria analysis in CARE-S
(2) Making plans for CCTV pipe inspection
(3) Risk analysis of flooding or pollution due to blockages

The tool can be operated as a stand-alone tool or through the Rehab Manager with import and export of data and results.

Data requirements

Input

The blockage tool requires three different input files:

(1) The pipe segments file, which contains data of all pipes that should be analyzed. If the tool is run through the Rehab Manager, the Rehab Manager generates this file, and the tool will only utilise information that is already present in the database. If the tool is run as stand alone, the file must be created and edited manually.
(2) The pipe blockages file, where two situations are possible:
a) the user has historic blockage data, and the tool analyses this data
b) the user has no blockage data, which means that the factors used for determining blockage probability must be manually entered instead of being calculated by the program.
(3) The pipe segment classification file defines how the blockage tool shall analyse the data. A default file is stored in the rehab manager, but it can/should be edited by the user to include groups of pipe characteristics (based on available data) or define ranges. This makes the blockage tool quite flexible, as the user can include several different types of information in the analysis, primarily in 'Blockage modus' or 'Root penetration modus'.

Table 2.2 Input files to Blockage tool

"Blockage modus":

Data	Unit	Kind of data	Data collection
Pipe Material Size/Diameter Age/Year of construction Pipe type	 mm	Pipe data base (basic information already present in Rehab manager) - modelling results (- measurements)	- Municipality
Blockages	#	Date of blockage and pipe ID	- Municipality

"Roots penetration modus":

Data	Unit	Kind of data	Data collection
Pipe Material Size/Diameter Age/Year of construction Pipe type	 mm	Pipe data base (basic information already present in Rehab manager) - modelling results (- measurements)	- Municipality
Pipe Depth	m	- manhole invert depth data	- Municipality
Soil Type		- maps (printed or digital) - point measurements associated to areas by geometrical measures - printed maps can digitised by using GIS programs	- Municipality, environmental agency, geological authorities - Land surveying office
Blockages	#	Date of blockage and pipe ID	- Municipality

Output

The output from the tool is a factor that indicates the probability of blockage for each individual pipe that has been assessed.

2.6 LOAD MODEL

General description

The Load Model is a computer program for gravity sewer pipes that calculates stress values in rigid non-reinforced concrete pipes based on external loads and pipe characteristics. The relation between load and resistance results in a security factor that is translated into reliability for structural failure. Nomenclature and calculation methods used in this model are based on the European standard CEN EN 1295-1.

Scope of application

A pipe will structurally collapse when its resistance gets lower than the stress inducing external and internal loads. The resistance is based on pipe material, diameter and wall thickness. External loads that are considered in the model are soil, traffic, groundwater and additional static surface loads. Frost load is not included since its effect is very dependent on further factors. Internal loads are water volume and pressure gradients. Those loads are under normal conditions reducing the stress in the pipe wall. The simplified calculation in the Load Model is therefore based on empty pipes. The resulting stress is calculated in circumferential and longitudinal direction. Besides pipe characteristic it is dependent on bedding conditions, side support and unsupported length caused by external soil erosion.

The pipe strength calculation will be based on the initial pipe strength at the time of installation. Combined with the loads, each pipe will have an initial risk factor for structural failure. The weakening factors over time are internal and external corrosion that will reduce wall thickness, weakening of pipe material caused by aging (leaching) and erosion of surrounding filling material that will increase the unsupported length. Two other models under CARE-S are calculating the external and internal corrosion of a pipe and are an essential input for the Load Model. Information from CCTV inspections can be used alternatively or additionally to those models. The reduced wall thickness will then result in a new risk factor for the prediction year chosen by the end user.

Strengths and limitations

The tool is designed to take into account the probable data availability at the end user site for sewer assets. The aim is to provide a tool that gives a qualified answer with a minimum of data demands, but the flexibility to model varying conditions. Assumptions, simplifications and adjustment possibilities in the load bearing calculation are described more detailed in the general model description.

Minimum input data requirements

For each pipe the following data is required:

- Pipe identification
- Diameter [mm] (is derived from shape, width and height. Width is equal to diameter)
- Installation year
- Wall thickness [mm]
- Installation depth [m]

Additional input data

- Traffic load (additional load from heavy traffic, can also be edited under calculating rules)
- External corrosion [mm/year] (can also be edited under calculating rules)
- Internal corrosion [mm/year] (can also be edited under calculating rules)
- Ground water level (equivalent to pipe installation depth, measured from surface)

In case missing wall thickness values, estimations can be done using SDR (standard dimension ratio) or PN (pressure class, only pumping lines).

Further data that is not provided by an input file can be changed under in the program itself:

- Unsupported pipe length (sensitivity analysis for longitudinal stress failure)
- Compaction condition (Load deflection in filling material)
- Bedding condition (3 standard types of embedment)
- Filling material [kN/m^3] (specific weight)
- Traffic load table (if necessary, the default values can be changed, see mathematical description)
- Pipe design strength [MPa] (manufactures value from load test)

Additional input data have already been assigned default values.

2.7 EXTCORR

General description

ExtCorr is a computer program that calculates external corrosion for concrete pipes.

Scope of application

The external corrosion is based on this equation:

$$C_{ext} = a * b\,(\text{year}_{\text{prediction}} - \text{year}_{\text{installation}})$$

Where the coefficients a and b are defined as:

a	=	external corrosion in mm per year
b	=	pipe quality (water cement ratio)

Coefficient *a* is chosen based on moisture grade and soil type, three different grades of each makes a total of 9 values. Suggestions for these values are provided. The moisture grade is defined as:

dry	when the pipe is more than 0,5 m above ground water level
medium	when the pipe is between 0 and 0,5 m above ground water level
wet	when the pipe is below ground water level

Coefficient b is chosen based on resistance grade. Resistance grade is based on installation year, suggestions are provided.

Minimum input data requirements

For each pipe these data are required:

- pipe identification
- diameter
- installation year
- wall thickness

Additional data:

- installation depth
- soil
- ground water level

In case soil information is missing, the user can input a specific year from when draining material is used as filling material.

In case wall thickness is missing, estimations can be done using SDR (standard dimension ratio) or PN (pressure class).

2.8 WATS INTERNAL CORROSION MODEL

General description

Microbial and chemical processes related to human waste have always caused problems and nuisances. Spreading of diseases, odors and unaesthetic conditions have been key concerns leading to the installation of sewer networks. By conveying waterborne waste in channels or pipes, odor and health problems are reduced significantly. However, confining the human waste to underground structures does not prevent microbial and chemical transformations to take place; it only constricts the processes to locations where they cause less harm.

Sewer corrosion

One problem that originates from conveying wastewater in sewers is hydrogen sulfide induced corrosion. Internal corrosion of sewers is mainly caused by microbial oxidation of hydrogen sulfide, which is produced under anaerobic conditions. Other internal corrosion processes of practical importance are due to discharge of chemicals into sewers. Such stochastic events cannot be modeled deterministically, and are consequently not included in the WATS model.

Description of the internal sewer corrosion model

The WATS model is a two-phase model – including wastewater and sewer atmosphere – solving a number of coupled differential equations in the water phase and in the gas phase of sewers, including transport as well as transformations. It simulates microbial and chemical transformation processes of organic matter, oxygen, and sulfurous compounds according to 2.3.

Table 2.3 Processes simulated by the WATS model

Phase	Transport and transformation process
Above the water table	Gas flow along the sewer line. Oxidation of hydrogen sulfide on the moist surfaces of the sewer walls.
At the gas/water interface	Transport of oxygen from the sewer atmosphere into the bulk water. Transport of hydrogen sulfide from the bulk water into the sewer atmosphere.
Below the water table and in pressure mains	Water flow along the sewer line. Transformation processes of organic matter, oxygen, and sulfurous compounds in the bulk water, the sewer biofilms, and the sewer sediments.

The model needs exact geometric information on the sewer layout, similar to what is needed in a hydrodynamic model. The calculation times using a Pentium 4 computer for e.g. 20.000 m of pipe are in the range of 0.5-5 seconds, depending on the sewer configuration.

The model calculates corrosion of individual pipes that are connected by manholes/nodes. It is basically a hydraulic model (for both gas and water phases) to which wastewater quality transformations and exchange processes between phases are coupled. The input for the model is the physical geometry of the sewer (slope, diameter, and so on) as well as dry weather flow inputs.

A large number of experimental data from laboratory, pilot and field studies constitutes the backbone of the model. The model is calibrated and validated under field conditions, resulting in default values for model parameters and wastewater components.

Scope of application

The processes taking place in sewers during conveyance of wastewater are of physical, chemical and biological nature. Physical processes are associated with the build up and erosion of sewer sediments. Chemical and physico-chemical processes are mainly gas transfer across phase boundaries (reaeration and emission of hydrogen sulfide and VOC's) but also hydrogen sulfide becomes chemically oxidized. Biological processes are the transformations of wastewater compounds associated with growth and maintenance of biomass.

Transformations of wastewater in sewer networks are related to processes in the wastewater phase, the slime layer (biofilm) and the sewer sediments. These biological transformations change the quality of the wastewater important for sewer corrosion, odor problems, wastewater treatment and combined sewer overflows. The transformations going on in sewers are strongly interlinked, and the importance of a process cannot be predicted without knowing the importance of all the other processes. As an example, to the question if hydrogen sulfide will occur in a certain gravity sewer, it must first be determined if the conditions are anaerobic (absence of both oxygen and nitrate). To do so, the complete mass balance for oxygen must be established, i.e. reaeration and oxygen consumptions in bulk water and biofilm must be known. In order to predict reaeration, the sewer geometry, temperature and flow conditions must be known, and in order to know the oxygen consumptions in bulk water and biofilms, the quality (composition of the COD) of the wastewater must be defined.

The simulation of corrosion and of odor is related to anaerobic conditions (no oxygen and no nitrate is present). Such conditions allow the biological processes like sulfate reduction and fermentation to proceed, which are causing corrosion and odor. Often these problems are observed under the following conditions:

- **After transport in pressure mains:** When wastewater is transported in pressure mains, anaerobic conditions rapidly will develop as no reaeration takes place. Even at low temperatures, significant H_2S and odor will be produced at typical transport times and COD concentrations. Due to volatile substances (like H_2S) being stripped off the water phase, corrosion and odor problems occur downstream of the pressure main outlet, especially where high turbulence exists.
- **In gravity sewers:** Corrosion and odor are seen in some gravity sewer systems. Conditions that increase the risk of such problems are: High temperatures, low pH, low flow velocities, large water depths, large biofilm to bulk water ratio, high COD and BOD concentrations, sediment deposits, and stagnant wastewater in e.g. septic tanks. Furthermore, if such wastewater is subject to high turbulence, stripping of hydrogen sulfide and other odorous compounds take place, potentially resulting in corrosion and odor problems.

The prediction of the different integrated in-sewer processes calls for a simulation tool – a computer model. The WATS model (Wastewater Aerobic/anaerobic Transformations in Sewers) is a dynamic model taking all the important biological and chemical processes in sewers into account. The model can yield information on e.g. wastewater quality, hydrogen sulfide and corrosion rates for any combination of time, space and network configuration.

The WATS model

The WATS model is a deterministic in-sewer process model for simulation of organic matter and sulfur transformations. The model consists of a number of non-linear differential equations describing the transformation of a number of different wastewater compounds. A software version of the WATS model has been developed with the purpose of dynamic simulation of biological corrosion of branched sewer networks.

What does the WATS model simulate?

The WATS model is a deterministic model dealing with transformations taking place in gravity sewers as well as pressure mains.

- *Biological transformations of wastewater COD.* The organic matter (COD) undergoes changes during transport. How and to what extent these changes take place depends on a number of conditions related to the layout of the sewer and the composition of the incoming wastewater. Transformations under aerobic and anaerobic conditions are covered.

- *Biological and chemical transformations of sulfur compounds.* When conditions become septic – e.g. when wastewater is pumped or conveyed in gravity sewers with low reaeration – H_2S is formed. When the H_2S is subjected to oxygen, it will be oxidized biologically as well as chemically. Oxidation takes place in bulk water and biofilm, but also on moist surfaces above the water surface. When oxidized on moist surfaces of the sewer, sulfuric acid is the end product, corroding concrete and metal.
- *Gas transfer at the water/gas interface.* In gravity sewers, the turbulence of the flowing water causes a continuous supply of oxygen from the sewer atmosphere into the wastewater. If wastewater becomes septic or not is determined by the balance between the oxygen supply and the oxygen consumptions of the water phase, biofilm and sediments. H_2S becomes a problem when released to the sewer atmosphere, where it can cause corrosion of moist surfaces.
- *Simulation of transport and transformation in both the water phase and the gas phase.* Transformations in the water phase take place in bulk water, biofilm and sewer sediments, while transformations in the gas phase only take place on the moist sewer surfaces. Bulk water and gas is transported with different transport velocities within the sewer. Furthermore, gas is lost by ventilation into the urban environment.

WATS focuses on dry weather problems and not wet weather problems. It deals with microbial and chemical transformation processes.

How does the WATS model simulate sewer processes?

WATS simulates microbial and chemical transformation processes of organic matter, oxygen, oxidized nitrogen compounds, and sulfurous compounds (2.5).

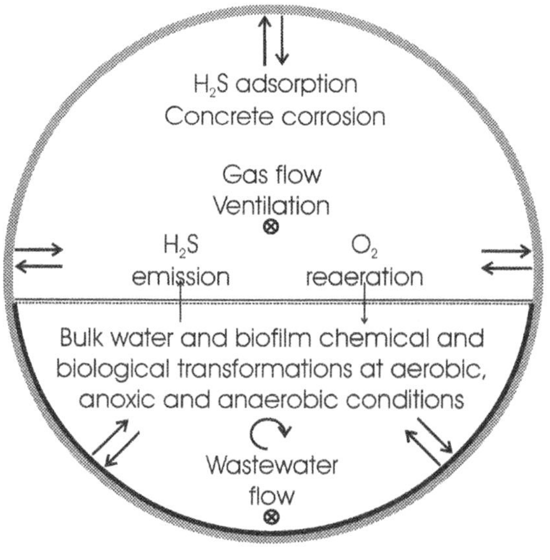

Figure 2.5 Microbial and chemical processes in a gravity sewer

Aerobic conditions in the water phase

When oxygen is present, heterotrophic microorganisms cause the main part of the oxygen consumption. They oxidize organic matter (COD) to yield energy for growth and maintenance. At the same time, the organisms use the COD as constituents in the formation of new biomass. Most of the COD present in wastewater is not immediately suited as substrate for the biomass. The molecules are typically too large to pass the cell walls and must be broken down into smaller, more readily degradable compounds. This process is called hydrolysis, and hereby the main bulk of the wastewater COD can be made available for the biomass. There is, however, always a small fraction of COD that is inert and cannot be made available for the biomass. This fraction is included in the slow hydrolysable substrate fraction (Figure 2.6).

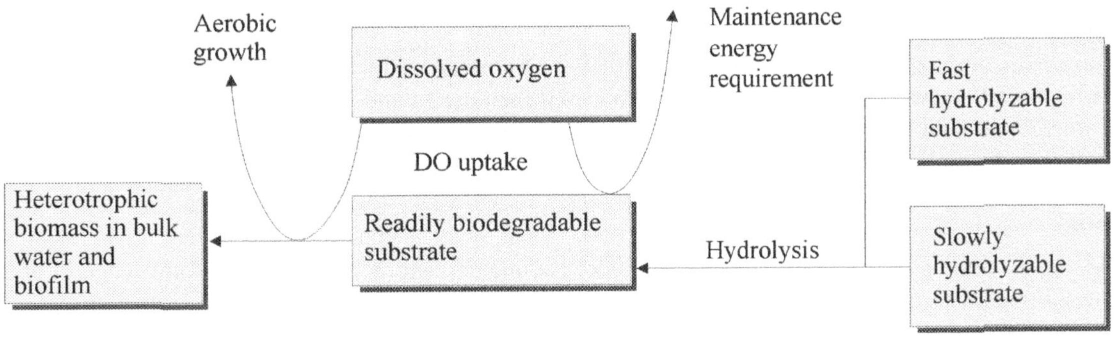

Figure 2.6 Aerobic transformation of organic matter

Hydrogen sulfide formed under anaerobic conditions becomes oxidized under aerobic conditions. The oxidation process is chemical as well as biological. Both process rates are significant under sewer conditions. The oxidation products are a mixture of oxidized sulfur forms, containing but sulfate and sulfur.

Anaerobic conditions

In the absence of both oxygen and nitrate, the previously mentioned heterotrophic organisms are no longer active. Instead other organisms take over. Some of these organisms ferment organic matter. Others reduce sulfate and oxidize COD. Another group of microorganisms produce methane from the COD. The first process – fermentation – gives rise to end products that are both desirable and undesirable. The end products have typically an unpleasant odor, however, they are excellent substrate for aerobic and anoxic treatment plant organisms. The second group – the sulfate reducing organisms – produce sulfide. Sulfides being related to a number of problems like odors, health hazards, toxicity towards microorganisms and corrosion.

Reaeration and hydrogen sulfide release

The two processes are strongly related, and can be described by the flow characteristics of the water phase. Both processes are relatively slow and consequently often limiting for the transformations in the sewer.

Corrosion

The hydrogen sulfide that is released into the gas phase of a sewer will be oxidized on the moist surfaces within the sewer. The end product of this oxidation is sulfuric acid, which attacks both concrete and metals and causes corrosion.

Rooting of gas and water

The rooting of water is simulated as stationary flow without dispersion. The reason for not using the full Saint-Venant's equations is that the WATS model simulates dry weather conditions. Here flow variations occur slowly and backwater is seldom an issue. Conveyance of gas is also modeled as stationary flow.

Input data

For the WATS model as well as any other model, it must be kept in mind that *The Model Output will Never be Better than the Model Input*. If only limited data can be made available, experience and knowledge on wastewater composition depending on catchment characteristics must be applied as a substitute and the accuracy of the prediction of corrosion decreases correspondingly.

It is seldom – or more correctly: it is never – possible to determine all inputs and all process parameters for a given sewer. This is partly due to the amount of information needed being rather large, partly due to a large natural variability of wastewater quality and process parameters, and partly due to the fact that we often want to simulate future scenarios. The WATS model must consequently be seen as simulating an "average situation", depending on the chosen system characteristics.

The WATS model is a deterministic model. It has a level of complexity similar to the hydrodynamic models used for simulation of stormwater runoff. For the pipe hydraulics, WATS needs the same data as any hydrodynamic model. In addition hereto, WATS must receive input data on the quality of the wastewater.

Strengths

In-sewer processes are manifold and complexly interwoven. Numerous transformation processes take place in the compartments of a sewer, i.e. in the bulk water, biofilms, sediments and gas phase, and different transport processes interconnect the various compartments. The conceptual WATS model addresses these aspects and allows a conceptual simulation of in-sewer processes and concrete corrosion.

Limitations

The WATS model contains a very large number of model parameters, which call for expert knowledge in order to calibrate the model to a sewer system. This process is complex and calls for an expert user with detailed knowledge on in-sewer processes and on the WATS model. The WATS model consequently comes with pre-set parameter values and without the possibility of calibrating model parameters. Where a detailed analysis including a calibration of the WATS model is needed, the user must make use of experts on WATS and in-sewer processes.

2.9 Z INTERNAL CORROSION MODEL

General description

The Z-model is developed for rough estimation of the risk of hydrogen sulfide related problems in sewer systems. It is valid for gravity sewers only.

Development of the Z-model

The gravity sewer assessment is based on a modification of the Z-formula, which is presented by Pomeroy (1970) and later applied by e.g. ASCE (1989), Melbourne and Metropolitan Board of Works (1989) and ASCE and WPCF (1982). The original formula is

$$Z = BOD_5 \frac{P/b}{s^{0.5}(35.314Q)^{0.33}} 1.07^{T-20}$$

Where:

BOD_5	=	biological oxygen demand during 5 days ($gO_2\ m^{-3}$)
P	=	wetted pipe-wall perimeter (m)
T	=	temperature (°C)
s	=	slope ($m\ m^{-1}$)
Q	=	flow ($m^3\ s^{-1}$)
b	=	pipe width at water surface (m)

According to recent studies on hydrogen sulfide generation (Hvitved-Jacobsen et al., 1998; 1999; Tanaka et al., 2000; Tanaka and Hvitved-Jacobsen, 2001), release (Yongsiri et al., 2003; 2004) and oxidation (Nielsen et al., 2003a; 2003b), it is deemed important also to include the biodegradability of the organic matter to a greater detail than what BOD_5 alone does, as well as including the pH into the equation.

Further information on the biodegradability of BOD can be obtained by the BOD/COD ratio. A high ratio means that the wastewater is relatively faster degradable and a low ratio means that it is relatively slower degradable. According to Henze et al. (2002) the average ratio between BOD and COD is 0.47. Values of 0.6 can be considered as high and 0.4 can be considered as low. The following equation consequently expresses the relative increase or decrease in the degradability of the BOD:

Where a value for ψ of 10 is estimated reasonable.

The modified Z-formula hence becomes:

$$Z'' = Z'\left(1 + 10\left(\frac{BOD_5}{COD} - 0.47\right)\right)$$

The dependency on wastewater pH is related to the dissociation of H_2S. Hydrogen sulfide is a weak acid with a pKa value of around 7, i.e. at a pH of 7, 50% exists as H_2S and 50% exists as HS^-. Only H_2S can be released to the atmosphere, and the lower the pH, the higher the release rate consequently becomes. E.g. at a low pH (lower than 5-5.5), the release rate is twice as high as at pH 7. On the other hand, at a high pH (higher than 8.5-9), the release rates approaches zero, and sulfide corrosion in sewers decreases considerably. Furthermore, the oxidation of H_2S in bulk water and biofilm is also dependent on which sulfide species are present: HS^- is significantly faster oxidized compared to H_2S. This fact increases the importance of the pH dependency, namely the fact that hydrogen sulfide problems are much more pronounced at low pH compared to high pH (Hvitved-Jacobsen, 2002).

The dissociation of H_2S is described by:

$$\frac{[H_2S]}{[H_2S_{tot}]} = \frac{1}{1+10^{pH-pKa}}$$

Because a typical pH of wastewater is between 7 and 8, it must be assumed that the Z-formula originally was developed for a pH higher than 7, and a pH of 7.5 being a reasonable assumption for the average of pH values. Assuming that the original study was performed at an average pH of 7.5, the correction factor for the original Z-formula becomes:

$$\left(1+10^{7.5-7}\right)\frac{1}{1+10^{pH-pKa}} \cong 4\frac{1}{1+10^{pH-pKa}}$$

The modified Z-formula hence becomes:

$$Z''' = Z''4\frac{1}{1+10^{pH-pKa}}$$

Combining all modifications, biodegradability and pH, the Z-formula becomes:

$$Z' = 4\left(1+10\left(\frac{BOD_5}{COD}-0.47\right)\right)\frac{1}{1+10^{pH-7}}BOD_5\frac{P/b}{s^{0.5}(35.314Q)^{0.33}}1.07^{T-20}$$

Scope of application

The application is to be used on gravity sewers only, and gives an estimate of the risk of the occurrence of anaerobic conditions, and hence the formation of H_2S, which will cause corrosion.

According to Pomeroy (1970), Z-values can be interpreted according to:

Z-formula	Estimated magnitude of sulfide problem
Z<5,000	Low risk of sulfide problem
5,000<Z<10,000	Some risk of sulfide problem
Z>10,000	High risk of sulfide problem

In the modified Z-model provided by the Rehabilitation Manager, the user can set these intervals according to local condition and experiences. It is recommended to do so, as the Z-formula originally was calibrated on a specific sewer network, and the calibration hence not necessarily valid for all systems.

Data

When choosing the parameters for the Z-formula, sewer geometry, flow regime, BOD, COD, temperature and pH has significant impact on the outcome of the prediction.

BOD and COD values depend on catchment characteristics like infiltration, industries, transport time, transport conditions and etceteras. The Z-formula is highly dependent on these values and it is consequently necessary to obtain good data. Note that BOD and COD values are not the same all over a catchment, but varies depending on catchment characteristics. Especially knowledge on industries is important in this respect.

When choosing the temperature, the purpose of the simulation must be considered. Analyzing for extreme situations with respect to odor and the toxicity of hydrogen sulfide gas, the yearly maximum temperature is a relevant parameter. However, when the purpose of the simulation is prediction of corrosion risk, choosing the average temperature is the better approach.

pH has a very pronounced effect on the dissociation of hydrogen sulfide, and – as only molecular hydrogen sulfide can be released to the sewer atmosphere – pH also has a significant effect on the resulting odor and corrosion problems. A typical pH of wastewater is in the range of 7-8, but both lower and higher values are not uncommon. pH as low as 6.5 in water with low alkalinity is not unseen, as are pH values as high as 8.5 for waters with high alkalinity. To simulate a catchment, pH values should therefore be obtained.

Values for BOD, COD, temperature and pH are commonly available from routine measurements on the wastewater quality to the inlet of treatment plants. However, variations throughout the catchment should be identified and taken into account.

Strengths

The Z-model gives a quick risk estimate on the risk of anaerobic conditions in the sewer system, and hence the risk of H_2S induced corrosion as well as odor problems. It is easy to use and requires a limited amount of data.

Limitations

The Z-model basically predicts the risk of anaerobic conditions, however, not the duration of anaerobic conditions. As H_2S formation is a rather slow, biological process, the duration of the anaerobic conditions is crucial. When analyzing the output of the model, this should be kept in mind. For example, if only a small part of the pipes in the catchment show high Z-values, this will not lead to corrosion, as the duration of the anaerobic condition is to short to allow H_2S to be formed. As a rule of thump, the duration of anaerobic conditions (i.e. the transport time in pipes with a high Z-value) should be above 1 hour before severe risk can be expected.

REFERENCES

ASCE (American Society of Civil Engineers) (1989). Sulfide in wastewater collection and treatment systems. ASCE manuals and reports on engineering practice no. 69, p. 324.

ASCE (American Society of Civil Engineers) and WPCF (Water Pollution Control Federation) (1982). Gravity sanitary sewer design and construction. ASCE manuals and reports on engineering practice no. 60 or WPCF manual of practice no. FD-5, p. 275.

Brillet F., 1995. Construction de lois d'évolution des chaussées par la méthode des lois de survie. Bulletin de Liaison des Laboratoires des Ponts et Chaussées, No 197, mai-juin, 43-53.

Curtis A., Molnar G. S., 1997. A municipal infrastructure management systems model. Canadian Journal of Civil Engineering, Vol. 24, CNRC, 1040-1049.

DeStefano P. D., Grivas D. A., 1998. Method for Estimating Transition Probability in Bridge Deterioration Models. Journal of Infrastructure Systems, Vol. 4, No. 2, June, ASCE, pp. 56-62.

Henze, M, Harremoës, P, Jansen, JC and Arvin, E (2002). Wastewater Treatment – Biological and Chemical Processes. Springer, 3rd edition, 2002.

Hvitved-Jacobsen, T., J. Vollertsen and N. Tanaka (1998). Wastewater Quality Changes during Transport in Sewers - an Integrated Aerobic and Anaerobic Model Concept for Carbon and Sulfur Microbial Transformations. Water Science & Technology, 38(10), 257-264. Printed in errata of 39(2).

Hvitved-Jacobsen, T., J. Vollertsen and N. Tanaka (1999). An integrated aerobic/anaerobic approach for prediction of sulfide formation in sewers. Water Science & Technology, 41(6), 107-116.

Hvitved-Jacobsen, T. (2002). Sewer Processes – microbial and chemical process engineering of sewer networks. CRC Press, pp. 237. ISBN 1-56676-926-4.

Li N., Haas R., Xie W.-C., 1997. Development of a new asphalt pavement performance prediction model. Canadian Journal of Civil Engineering, Vol. 24, CNRC, 547-559.

Madanat S., Mishalani R., Wan Ibrahim W. H., 1995. Estimation of Infrastructure Transition Probabilities from Condition Rating Data. Journal of Infrastructure Systems, Vol. 1, No. 2, June, ASCE, 120-125.

Mailhot A., Duchesne S., Musso E., Villeneuve J.-P., 2000. Modélisation de l'èvolution de l'éttat structural des réseaux d'égout : application une municipalité du Québec. Canadian Journal of Civil Engineering, Vol. 27, CNRC, 65-72.

Melbourne and Metropolitan Board of Works (1989). Hydrogen sulfide control manual – septicity, corrosion and odour control in sewerage systems. Technological Standing Committee on Hydrogen Sulfide Corrosion in Sewer Networks, vols. 1 and 2.

Nielsen A H, Vollertsen J and Hvitved-Jacobsen T (2003a). Determination of Kinetics and Stoichiometry of Chemical Sulfide Oxidation in Wastewater of Sewer Networks. Environmental Sciences & Technology, 37(17), 3853-3858.

Nielsen, A.H., Vollertsen, J. and Hvitved-Jacobsen, T. (2003b). Chemical sulfide oxidation in wastewater – effects of pH and temperature. Proceedings from the 2nd IWA International Conference on Odors & VOCs, Singapore, 14-17 September 2003.

Pomeroy (1970). Sanitary sewer design for hydrogen sulfide control. Public Works, October, 93-96 and 130.

Tanaka, N., T. Hvitved-Jacobsen and T. Horie (2000). Transformations of carbon and sulfur wastewater components under aerobic-anaerobic transient conditions in sewer systems, Water Environment Research, 72(6), 651-664.

Tanaka, N. and T. Hvitved-Jacobsen (2001), Sulfide production and wastewater quality – investigations in a pilot plant pressure sewer, Water Science & Technology, vol. 43, no. 5, 129-136.

Yongsiri C, Vollertsen J and Hvitved-Jacobsen T (2003). Hydrogen sulfide emission in sewer networks: a two-phase modeling approach to the sulfur cycle. Proceedings from the 2nd IWA International Conference on Odors & VOCs, Singapore, 14-17 September 2003.

Yongsiri C, Vollertsen J and Hvitved-Jacobsen T (2004). Air-water transfer of hydrogen sulfide – an approach for application in sewer networks. Water Environment Research, 76(1), 1-8.

WRc, 1994. Sewerage Rehabilitation Manual (3rd ed.)

3

Description and validation of hydraulic performance

Rita Ugarelli, University of Bologna

INTRODUCTION

The performance of a sewer system can be considered as its ability to transport storm and wastewater without hydraulic overload, as well as creating minimal environmental impact and retaining good structural integrity. This can be achieved by judging performance against the following criteria: water quantity standards, maximum water levels, allowable discharges, water quality standards, overflow frequency and structural condition standards.

This report describes the contents of the CARE-S Work Package 3 (WP3) and presents the problems considered, methodologies developed and results reached.

The CARE-S Work Package 3, "Description and validation of hydraulic performance", deals with the study and understanding of the hydraulic behaviour of the sewer system affected by temporal decline, with the environmental impact of hydraulic deficiencies and with the final definition of the system reliability. The hydraulic performance is the most important parameter for the analysis of the level of service of the wastewater collection system. It depends on the dimension of the pipes but also on temporal effects of structural deterioration, blockages, roots, sags, collapses and sediment transport, which affect the capacity to transport wastewater and runoff. Aim is it to avoid local floods and excessive pollution discharges.

The sewer system is connected to the wastewater treatment plant (WWTP) and to the natural compartments groundwater and surface waters. Therefore, the urban wastewater system must be perceived as an integrated and complex system, where changes in one of the subsystems affect the conditions in all the other subsystems.

WP3 activity, planned in four tasks, started with the evaluation of available 1D software (commercial or freeware) capable of modelling water flow, water quality and sediment transport in sewer systems, receiving water bodies, WWTP and groundwater (task 3.1).

Hydraulic simulations were performed in an innovative way, by introducing temporal deterioration of the system as described below (task 3.2). CCTV inspections data, collected as recommended by the E.U. standard 13508-2:2001, have been used as input data for 3D simulations which is useful for the comprehension of failure hydraulic effects.

Task 3.3 aims at analysing environmental impacts of rehabilitation. A methodology for system analysis by using simplified approaches within CARE-S or integrated simulation outside of the CARE-S procedure using SIMBA® and SIMBASewer® models has been developed . A synchronous simulation of the sewer network, the wastewater treatment plant and the receiving water and their interactions is possible with this software package.

Reliability considerations for sewer and storm water networks are an integral part of all decisions regarding the planning, design and operation phases of such systems. The task 3.4 of WP3 is focused on the integrated performance of the whole sewer system and will form a basis for identifying potentially high-risk sewers.

3.1 WORK – PACKAGE 3.1: EVALUATION OF CURRENT BEST PRACTICE

3.1.1 Task and reports

The task is based on the evaluation of available software (commercial or freeware) capable of modelling water flow, water quality and sediment transport in sewer systems, receiving water bodies, wastewater treatment plants and groundwater.

Modelling is needed at different level of complexity: in integrated urban drainage modelling the adoption of detailed models leads to unacceptably long calculation times and implies large memory needs. Therefore simplified models are appropriate to very quickly simulate long-rainfall time-series, thus allowing the evaluation of the environmental impact on receiving water. However, detailed models are needed in order to evaluate discharge, depth, and velocity in each pipe of the sewer system and its ancillary structure (e.g. Combined Sewer Overflow).

The review activity, reported in Deliverable D7 (*Freni G. et Al, 2003*), conducted to the selection of three 1D models which will be included in the CARE-S Rehabilitation Manager Prototype: MOUSE (developed by Danish Hydraulic Institute), InfoWorks CS (developed by Wallingford Software) and SWMM (developed by U.S. EPA).

3.2 WORK – PACKAGE 3.2: MODELLING HYDRAULIC PERFORMANCE AND TEMPORAL DECLINE

3.2.1 Task and reports

The aims of task 3.2 can be summarised as follows:

(1) Describe hydraulic investigation and analysis;

(2) Extend the standard hydraulic modelling approach used in 1D Urban Drainage Models (UDM) to cover a temporal decline of the system's hydraulic performance by attributing hydraulic characterisation of failures, defined in compliance with the recently adopted European defect coding system EN13508-2 (k headloss coefficient, lambda of Colebrook formula and M, Manning coefficient) and computed using 3D modelling software (FLUENT);

(3) Provide the end-users of CARE-S Rehabilitation Manager with a possibility to assess hydraulic performance of the sewer system including temporal decline inside the simulations using the 1D UDM.

Good understanding of procedures and methods being used in the commercial 1D UDM software packages MOUSE, InfoWorks and SWMM has crucial importance in the development of the WP 3 concept especially for the method for hydraulic performance of temporal decline and reliability modelling: three CARE-S end-user case studies have been used to support the WP 3.2 concept development and reinforce its links with other WPs and CARE-S Rehabilitation Manager.

The three case studies on hydraulic performance modelling which are based on the MOUSE, INFOWORKS and SWMM software packages were established and calibrated by CARE-S end-users in Oslo (Norway), Correggio and Palermo (Italy).

The results are reported in Deliverable D8 (*Milina J. et Al., 2004*).

3.2.2 Hydraulic investigation

Hydraulic investigation and analysis is usually fundamental part of a sewer network performance evaluation. The purpose of this stage is to obtain an objective assessment of system performance. The extent of data collection and analysis involved will be determined by the information available and the rehabilitation needs.

Different forms of assessments require different types of data and level of detail. Where criteria based on continuous performance are to be assessed, measured data is usually the most appropriate. Where assessment covers performance associated with extreme events that occur infrequently, measured records will normally prove to be inadequate. In these circumstances, simulation of system performance using software-modeling tools supported by short-term data collection is often the only way forward. Table 3.1 summaries appropriate data and modeling for assessing different performance objectives. In addition, the performance of the system can be assessed qualitatively by visual inspection, directly or through CCTV analysis.

Table 3.1 Data collection methods and modelling approaches for assessing the hydraulic performance of sewerage systems.

Performance Objective	Data Collection and Modeling Approaches
The pipe work operates without blocking	CCTV records, manual inspections
Drains and sewers shall be water tight	Long term flow monitoring, short term flow survey
Odour, nuisance and toxicity do not arise	CCTV records, manual inspections
The sewer surcharge frequencies shall be limited to prescribed values	System modeling verified by short term flow survey, CCTV survey, manual inspections
The flooding frequencies shall be limited to prescribed values.	System modeling verified by short term flow survey
Public health and life shall be safeguarded.	System modeling verified by short term flow survey
Receiving waters shall be protected from pollution	System modeling verified by short term flow survey and quality sampling, CSO event monitors, aesthetic pollutant observations, macro-invertebrate surveys

3.2.3 Carry out flow measurement

It is beneficial to use data from as many different sources as possible when calibrating and verifying system models. There are generally three categories of data that can be used:

- Historic Data: flooding records, complaints' records, recorded pollution incidents.
- Survey Data: short-term flow monitoring, pollutant sampling, macro-invertebrate sampling, aesthetic pollutant surveys.
- Permanent Monitoring Data: sewer flow gauging stations (e.g. for RTC), treatment works flow records.

3.2.4 Evaluation of temporal decline effects on the flow capacity using the 3D model FLUENT

Sewer systems fail in a number of ways. Failures are generally categorized by the effects they have on their surrounding environment, rather than the condition of the system itself.

Two general failure categories for sewer systems can be defined:

- Technical pipe state or failure
- Structural collapse (due to physical or microbial/chemical deterioration)
- Operational pipe state or failure
- Infiltration of groundwater
- Exfiltration of sewage
- Sewer blockages or "chokes" (sediment build up, ragging, root intrusion)
- Design deficiencies 2 (e.g. negative slope, "sags", bottlenecks, etc.)

A model will be required for each of these failures.

Classification of performance and structural condition data should be carried out in accordance with EN 13508 (European Standard – prEN 13508-2 – Condition of Drain and Sewer systems outside buildings – Visual inspection coding system): as proposed in this standard, codes commencing "BA" related to the fabric of the sewer or drain, and codes commencing "BB" relate to the operation of the sewer and drain, i.e. its hydraulic and/or environmental functioning.

Some of the "BA" and "BB" codes have been analysed through 3D numerical analysis in order to evaluate the reduction of hydraulic capacity from visual inspection code.

Defect classification in accordance with EN 13508-2:2001

Internal inspections of sewers, either direct visual inspections in man-entry sewers, or Closed Circuit Television (CCTV) inspections of smaller sewers, will normally form an essential part of the planning of sewerage rehabilitation.

The actual inspection of manholes and sewers depends on the type of information required, i.e. structural condition, hydraulic deficiencies, environmental aspects or connectivity queries.

The coding system of EN 13508 is designed to enable defects to be described quickly and easily with a minimum of paperwork. Operatives and Engineers should be familiar with the coding system and work to the same one. Operators should be trained to use the equipment, recognise the relevant features/defects and be aware of the appropriate coding system(s).

The coding system is made with 3 letters and little additional information: the first letter describes the code application (does it refer to a pipe or a manhole or…?), the second deal with the kind of code (does it refer to material, usage, or…?), the third defines the failure (deformation, sags, crack, obstacle or…?); the additional information helps to quantify the observation or to describe the failure position.

EN 13508 form the basis of a systematic approach to identify hydraulic deficiencies such as reduction in sewer capacity due to an increase in pipe roughness according to the features from the visual inspection survey or due to a reduction in pipe cross – sectional area. Infiltration and exfiltration can be accounted by representing a reduction and increase in capacity respectively depending on the severity of the flow encountered.

3D numerical analysis using FLUENT model

Methodology

The hydraulic capacity of sewer systems affected by failures has been simulated using a mathematical 3D model, FLUENT. Each failure was simulated separately. Simulations were performed on fictitious pipes trying to represent all the possible failure conditions. The methodology is based on the comparison of pressure drops between a new clean pipe and a pipe deteriorated with the same characteristics and material. Failure effects have been calculated in term of roughness coefficient. Knowing the pipe deterioration level from CCTV inspection data, the calculation of the roughness coefficient can be performed and the received value is used as input to UDM model instead of the usual default values.

3D methodology and mathematical modelling applied are described in "*The hydraulic capacity of deteriorating sewer Systems*"(*Ugarelli R. et al., 2004*).

Starting from pressure losses between the beginning section and the end section of a fictitious pipe and between the same sections of a pipe with failure, it is possible to evaluate new values for the Manning "n" or Colebrook "λ" coefficient. Those values will be used to simulate the failure presence inside the pipe.

Head losses produced by failures can be simulated either as local head losses or as distributed head losses along the pipe. When obstacles inside the pipe are simulated with FLUENT, their effect is given as variability of the k coefficient of local head loss:

$$\Delta h = k\frac{v^2}{2g} \qquad (1)$$

The local head loss is difficult to simulate using the 1D UDM without changing the sewer network topology, but it is possible to convert the k coefficient in the Manning roughness coefficient n using the following equation:

$$n_{new} = \sqrt{\frac{10{,}29n^2}{10{,}29 + (\frac{0{,}082711D^{\frac{4}{3}}n^2k}{L})}} \qquad (2)$$

Simulation of obstacles

Obstacles were divided into groups and these groups were simulated separately. Failures were simulated as if they occurred in the middle of the pipe. From difference of pressure drop between the beginning and the end of the fictitious pipe and the pipe with failures, it is possible to evaluate the local head loss due to the failure.

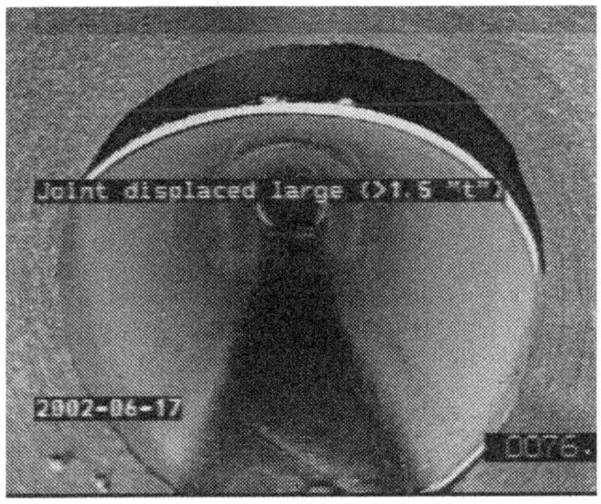

Figure 3.1 Example of displacement pipe scanned by CCTV inspection

Displaced pipe (EN 13508 code: BAJ)

This simulation of obstruction in sewer system was simulated as two pipes of 10 m in length with different levels of displacement connection. The example figure (Figure 3.2) shows pressure drop on 20 m length pipe due to two types of displacements: 20% and 10% of displacement. Depending on the percentage of displacement the final formula obtained to calculate the local head loss due to this kind of failure is:

$$k = 0{,}0122 \cdot e^{12{,}295 \cdot \%} \tag{3}$$

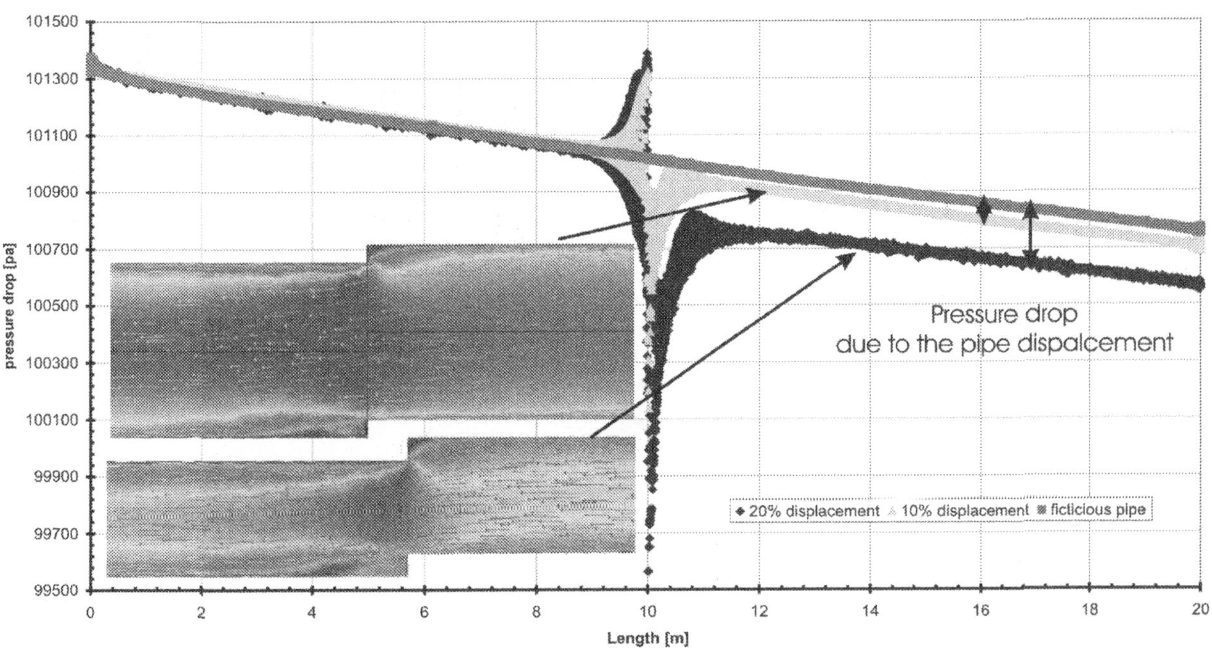

Figure 3.2 pressure losses due to pipe displacement

Roots (EN 13508 code: BBA)

Tree roots may grow into the pipe through cracks or joints. For the simulation of roots intruding into sewer a different model was developed. Roots were simulated as a porous media.

At high flow velocities, the constant C_2 in next equation provides a correction for inertial losses in the porous medium. This constant can be viewed as a loss coefficient per unit length along the flow direction, thereby allowing the pressure drop to be specified as a function of dynamic head. For modelling a perforated plate or tube bank, we can sometimes eliminate the permeability term and use the inertial loss term alone, yielding the following simplified form of the porous media equation:

$$\nabla p = -\sum_{j=1}^{3} C_{2_{ij}} \left(\frac{1}{2} \rho v_j v_{\max} \right) \tag{4}$$

Coefficient used in the roots model for inertial resistance: $\mathbf{C_2 = 5}$ in all the x, y, z directions.

The developed formula to simulate root intrusion effect is:

$$k = 2{,}3913 \cdot \%^2 + 0{,}0971 \cdot \% + 0{,}0889 \tag{5}$$

Obstacle "Partly intruding pipe" (EN 13508 code: BAG).

This is the simulation of pipe or cables intruding sewer. This model is used for intruding connection and for obstacle model. Several different sizes and distance of intrusion were simulated. Depending on the percentage of intersection area occupied by intruding pipe the final formula were obtained to calculate the local head loss:

$$k = e^{14{,}673 \cdot \%^3 - 17{,}544 \cdot \%^2 + 12{,}324 \cdot \% - 3{,}6518} \tag{6}$$

Sags (EN 13508 code: BBE)

Sags are caused by failure of the pipe bedding on the bottom of the trench. This type of failure causes a section of the pipe to drop below proper grade. Water remains trapped in the sag and solids suspended in the water tend to settle in the sag area.

$$k = 3 \cdot 10^{-5} \cdot \alpha^2 + 0{,}0047 \cdot \alpha + 0{,}0921 \tag{7}$$

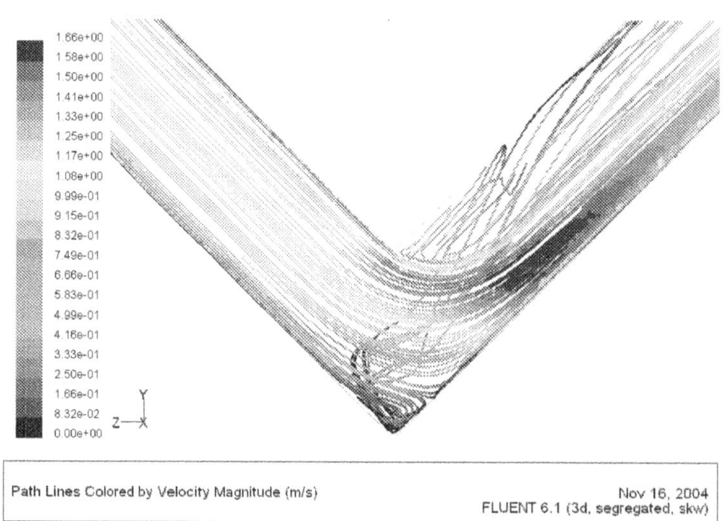

Figure 3.3 Path lines coloured by velocity in 90° elbow

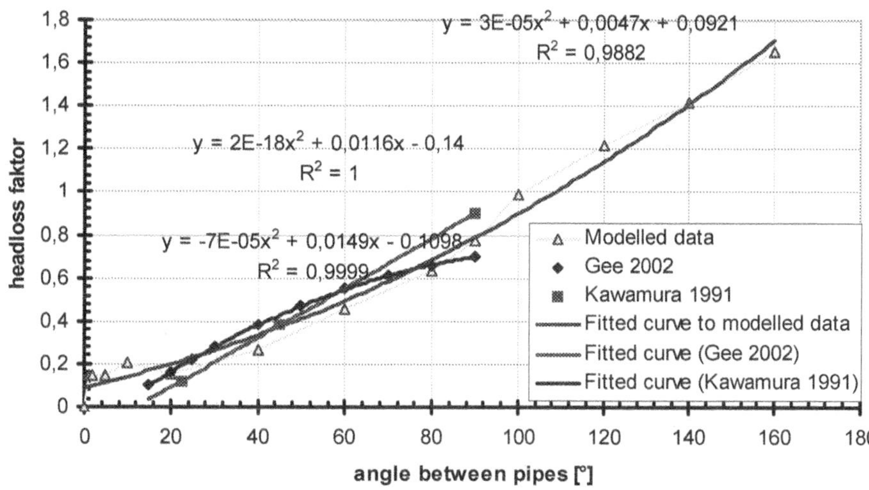

Figure 3.4 Sags pressure losses compared with literature

Study of sags was focused mainly to small angles (ά), because these are more obvious. However, larger angles were studied too to compare it with literature review. Difference between literature and modelled data are mainly due to different shapes of elbow. We are considering sag to be sharp, but in literature the elbow is not too sharp (elbow in water pipe).

Experiment

For the validation of result an experimental pipe was created. Experiments were performed in the Norwegian University of Science and Technologies.

The first experiment was run without any obstacle to compare pressure losses due to the wall friction. The second part of the experiment was performed with the insertion of obstacles in the middle of the experimental pipe. Two types of obstacles were inserted and compared with mathematical model. Pressure losses on experimental pipe were modeled. On the beginning of the curve you can see pressure drop due to the inflow. On zero is first pressure measurement and second pressure measurement is on the end. In the middle is pressure loss due to the obstacle.

Results from experiment show a good accordance with the modelled data. Data from experiment and FLUENT simulations for the same flow rate were compared. Only the comparison between pipes without obstacles shows difference between experiment and modelled data. It was due to small difference and fluctuating inflow rate. (*Pollert J, 2004*)

Results

Figure 3.5 shows the curves developed, equations to simulate the local head loss produced by different kind of failures: obstacles, pipe's displacement, roots and sags. Result shows very similar local losses for similar kinds of obstacles for identical intersection area occupied by obstacle.

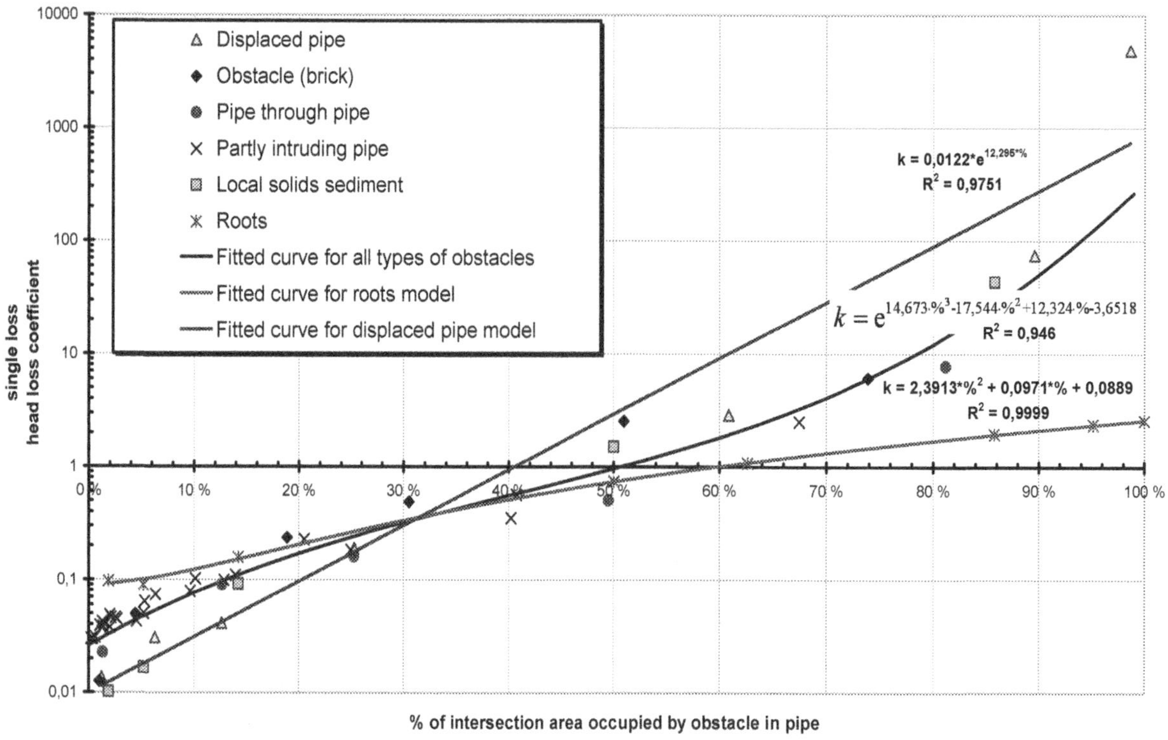

Figure 3.5 Overall results – several different types of simulated failures

3.2.5 TOOL: DEGRADATION MODEL

The 3D numerical analysis allowed the evaluation of sewer degradation effects on the flow capacity. The effects are established in term of pressure drops, local or distributed, calculated with the developed formulae as function of the pipe section reduction. In order to automatically load the headloss information to the 1D models as input a Visual Basic code has been developed, named Degradation tool.

The degradation tool received from the CARE-S prototype the input data:

- CCTV information for each pipe effected by failures;
- Pipe data (length, diameter, roughness coefficient, shape)

The visual basic code calculates, using the formulae developed with FLUENT, the new roughness coefficient for pipes depending on the specific level of degradation. As output for each degradated pipe we receive a new manning or a new lambda or a local headloss coefficient. The result is given in different ways because the 1D models are calculating the pressure drops with different coefficients.

The output file is automatically driven by the prototype to the 1D model where the degradation data become input for new hydraulic simulations.

Example of degradated system analysis: Nordberg case study

The 3D simulation results were tested and validated on the Nordberg sewer network (Oslo - Norway). The procedure which was carried out using an already established MOUSE model of the Nordberg sewer system in Oslo, involves three main steps:

- Identification and "quantification" of the classified failures in the Norwegian CCTV inspection system;
- Attribution of the hydraulic characterisation of failures (k, λ, M) computed using 3D modelling software (FLUENT) to the existing methods of 1D hydraulic simulator (MOUSE);
- Hydraulic performance modelling and reporting the effects of failures using surface flooding as performance indicator.

The results of the analysis are reported using surface flooding as performance indicator.

Figure 3.6 summarises the result of the simulation run on the Nordberg system. The results of the analysis are reported using surface flooding as performance indicator. The hydraulic consequences of the modelled failures are presented on the right side of Figure 3.5, while the hydraulic consequences of a "new" sewer system (without failures) are shown on the left side of Figure 3.5. While the new sewer system can cope with a rainfall event that is only exceeded in severity once in 5 years, the temporal decline of the "old" system leads to one localised surface-flooding incident. An event that is only exceeded in severity once in 20 years implies, in the "old" system, an increased number of pipelines causing surface flooding, as well as a greater impact of flooding incidents.

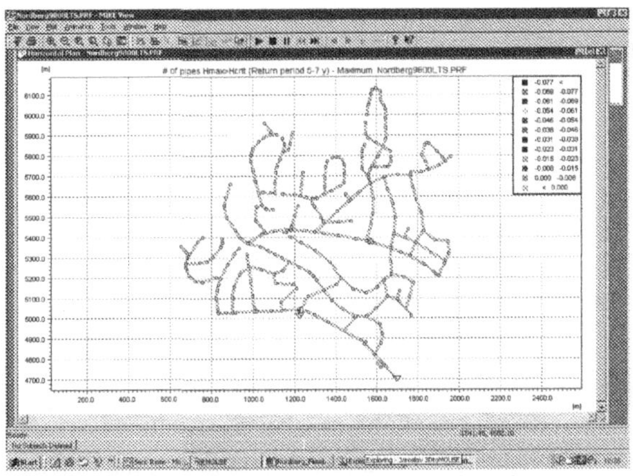

a1) Model without failures - there is no surface flooding.

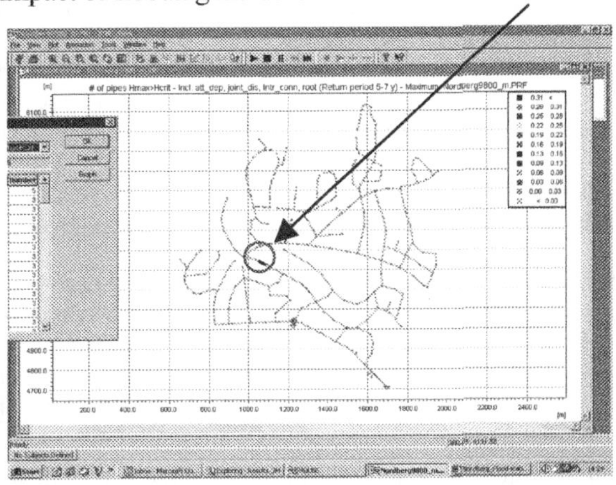

a2) Failures included into model - there is surface flooding.

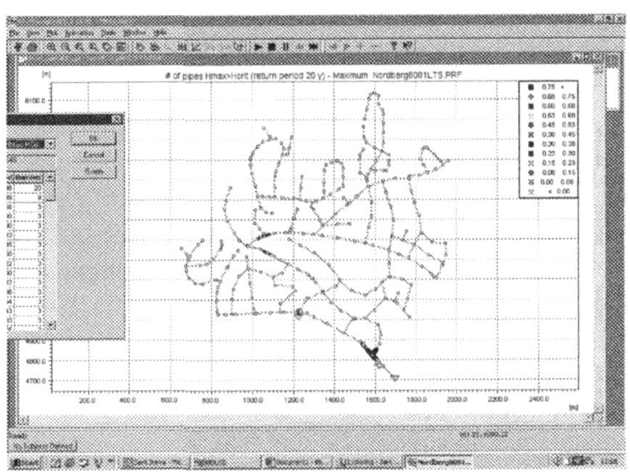

b1) Model without failures - there is surface flooding.

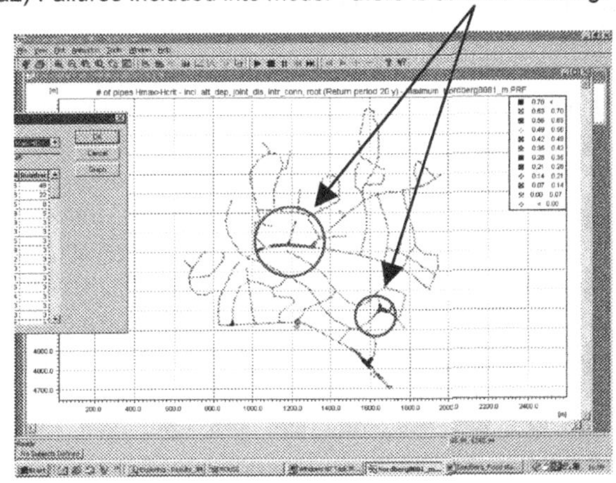

b2) Failures included into model - surface flooding is severer.

Figure 3.6 Effects of a set of failures on frequency of surface computed using 1D simulator (MOUSE). a) Events with 5 years return period, b) Events with 20 years return period

3.2.6 Identify hydraulic deficiencies

Once the hydraulic investigations have been completed, the CCTV data collected, the performance of the existing system can be objectively assessed. This is achieved using performance criteria and will normally involve a combination of measured data and simulated data from a simulation model.

This requires the model to be run with appropriate rainfall event data. Some modelling software tools have been written such that the frequency of model output (peak flow, volume of flow etc.) corresponds to the frequency of rainfall event input. The rainfall event data is chosen, therefore, to give the desired frequency of model output. For certain aspects of performance it will be sufficient to use single event data. This applies when assessing performance for the more extreme events. For example, the extent of surcharging or flooding that occurs with a five-year return period can be assessed using a five-year return period rainstorm. This exercise should be undertaken with rainfall profiles of differing durations and peakedness to ensure that the worst case is evaluated.

When evaluating the hydraulic performance of a sewerage system, it is important to record not only the locations where the performance falls below the agreed standard, but also to assess the degree of failure and the consequences of failure. This will enable the seriousness of any failure to be assessed. The assessment can then be used to determine investment patterns and evaluate the cost effectiveness of rehabilitation proposals. Table 3.2 indicates how the performance assessment might be structured.

Table 3.2 Example of hydraulic performance assessment

Performance Objective	Performance Target	Assessed Performance	Consequence of Failure	Priority for Rehabilitation
The pipe work operates without blocking	No evidence of blockage from CCTV surveys	Partial blockage observed in a small number of trunk sewers	Reduction in hydraulic capacity increasing frequency of CSO operation	Medium
Drains and sewers shall be water tight	Infiltration less than 50l/head/day	Treatment works records show infiltration at 200 l/head/day	Early and more frequent spill from CSOs, more rapid structural deterioration of sewers.	Medium
The sewer surcharge frequencies shall be limited to prescribed values	Sewers surcharge no more frequently than once per annum on average	Sewers heavily surcharged for one year return period event	More rapid structural deterioration of sewers	Low
The flooding frequencies shall be limited to prescribed values.	Property not subject to flood more frequently than once in thirty years on average	City centre commercial areas flooded on average once every five years	High cost of consequential loss to property and business	High

The link between cause and effect is an essential part of developing rehabilitation options. Often the location of a cause of performance deficiency is remote from the location of the effect. For example, property flooding may be due to the lack of capacity of the sewer pipes in the immediate vicinity of the property, or it may be due to the overloading of a trunk sewer downstream leading to the backing up and surcharging of the system adjacent to the property. If the cause is the latter, then local upsizing of the sewer pipes will prove fruitless.

Table 3.3 summarises the hydraulic issues for problems that may be caused directly hydraulic or environmental system failures.

Identification of problems will usually help to establish potential rehabilitation solutions. It should be appreciated however, that solutions development should not be undertaken in an un-structured way. Since changes to a system generally have a consequential effect downstream, rehabilitation options should always be evaluated from the upstream end of the system, working in a downstream direction. In selecting possible options, the following order is recommended:

- Localised repair and/or joint sealing to reduce in/exfiltration.
- Removal of local restrictions such as blockages or partial collapse.
- Removal of inputs by finding alternative disposal routes.
- Provision of above ground storage to limit inputs.
- Provision of below ground storage to attenuate flows.
- Re-setting of CSOs to divert more flow to receiving waters
- Upsizing sewers to convey increased flow

Table 3.3 Hydraulic issues and generic methods of deriving solutions

Possible causes	Possible solutions	Available methods
Lack of capacity	Reduce flows	Disconnect contributing areas
high infiltration		remove foul / industrial flow
illegal connections		remove infiltration
Build up of fat / grease		
Ingress of tree roots	Increase system capacity	add a relief sewer
Siltation		add storage
Increase in discharge		upsize pipeline
		clean sewer
	Attenuate flows	throttle inflows
Structural failure	Improve structural integrity	renovation sewer
pipe collapse		replace sewer
pipeline distortion		
Mechanical failure		
pumping station		
rising main		
flow control / RTC		

3.3 WORK-PACKAGE 3.3: ENVIRONMENTAL IMPACTS OF REHABILITATION

3.3.1 Tasks and reports

Work package 3.3 aimed at the consideration of environmental aspects in the CARE-S procedure. Tasks were the identification of problems and driving factors related to:

- Groundwater pollution and groundwater interaction with the sewer system (infiltration, exfiltration)
- Wastewater treatment plant performance due to changed wastewater composition
- Receiving water quality

The results are reported in Deliverable D9 (Schulz and Krebs, 2004). The task of developing a model for infiltration and exfiltration was shifted to CSIRO, where the colleagues were involved in the development of similar models for the AISUWRS project. Additionally to the work description it was decided during the second CARE-S meeting to provide computer programs to calculate environmental criteria for the decision support within the program CARE-S. The programs have been included in the CARE-S rehabilitation manager. On these programs also a report has been provided (Schulz, 2004).

During the second CARE-S meeting it was decided not to include quality models in the CARE-S procedure. A comprehensive analysis of environmental deficiencies would require on site measurements or pollution transport modelling in the sewer and the use of wastewater treatment plant models as well as receiving water models. Therefore, this extensive analysis is not part of the CARE-S procedure. Here screening methods and tools are provided to prioritise areas where detailed investigations or modelling studies should be carried out.

3.3.2 Problem identification

The sewer network interacts with groundwater, surrounding soil and surface waters depending on boundary conditions, which are, among other factors, the condition of the pipes. Infiltration of groundwater into the sewers might affect the performance of the wastewater treatment plant (WWTP) and the combined sewer overflows (CSO) in the system. Due to these effects oxygen deficits or toxic ammonia concentration in small receiving waters can be reinforced. Exfiltration of wastewater into the surrounding soil is a possible source of contamination of groundwater and should be prevented. The results of a questionnaire sent to the end-users of CARE-S confirmed that in many cases infiltration, exfiltration and wastewater treatment plant issues are the most important environmental aspects in rehabilitation planning. In the literature survey possible environmental effects related to these issues due to failures in the network condition or performance are described.

Table 3.4 Definition of failures related to the environment that are considered in CARE-S

	Description of level of performance via	Criteria	Value
CSO	PI target or trigger Standard/regulation Voluntary objective	Volume Spill frequency Load Design flow	Depends on national or local regulation
WWTP	Standard/regulation for inflow Voluntary objective Initial design		
Exfiltration	Any occurrence of exfiltration PI target or trigger	Vulnerability of specific area; Exfiltration volume	Yes/no, Low or high vulnerability
Infiltration*	PI target or trigger Any regulation or voluntary objective	Infiltration volume	User defined values
Receiving water	Standards and regulations Voluntary values	Discharged total load CSO criteria Hydraulic load	

*The level of performance required for infiltration depends mainly on the consequences the extraneous water flow has for CSO, WWTP or other parts of the urban drainage system. The decision whether infiltration is a failure or not can only be made by simultaneously assessing the other compartments of the system.

As supplement to the literature survey the influence of increased infiltration due to network deficiencies on the receiving water has been simulated exemplarily with a hypothetical system. Widely applied wastewater treatment plant models like the ASM1 (Henze *et al.*, 2000) reflected the findings from literature very well. The influence on combined sewer overflow performance, especially of pollutant transport, can also be modelled, but are subject to significant uncertainties in the model structure and in data assumptions.

3.3.3 Derivation of criteria to be used in the decision support system

Criteria were derived to include the environmental aspects in the decision support procedure. The criteria can be used to assess or rank rehabilitation projects, or to design or assess long-term rehabilitation strategies. Wastewater treatment plants are not directly considered in the decision support procedure. Alternatively a report on possible problems in wastewater treatment plants due to network deficiencies is given within the Deliverable D9.

Criteria are:

- Vulnerability of groundwater
- Combined sewer overflow performance (compliance to standards) in terms of total load, frequency/number of spills, volume and duration
- Criteria to estimate the pollution and hydraulic impact of the entire urban drainage system on rivers

The vulnerability of groundwater will be a criterion in the selection of rehabilitation projects and in the selection of a long-term rehabilitation strategy. Combined sewer overflow aspects will only be used in the long-term assessment. Figure 3.7 shows the procedure of integrating environmental modelling into the long-term assessment.

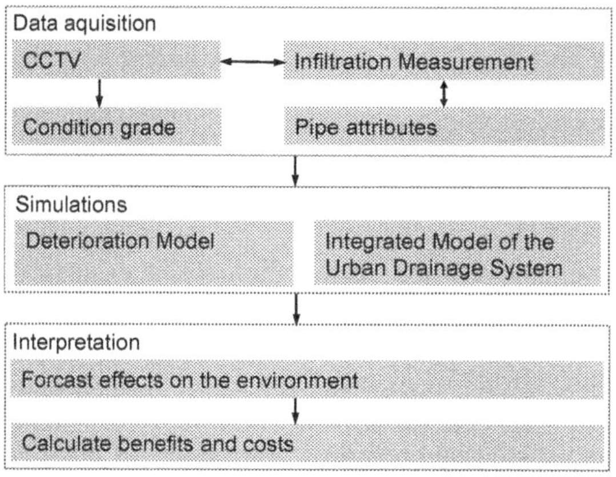

Figure 3.7 Flowchart of rehabilitation planning using urban drainage modelling

If various rehabilitation scenarios have been modelled using the hydraulic model and no optimum solution can be found a list of possible measures apart from network rehabilitation is provided which is included in the software. The list is a proposal to the user. It contains a number of problems and possible solutions in the network, the receiving water and in the catchment. The list is a database, which aims at the development of alternative solutions outside of the CARE-S procedure.

3.3.4 Methods to calculate criteria

Vulnerability of groundwater

The pollution of groundwater in urban areas due to leaking sewers has been investigated and confirmed by several studies. However little is published on health impacts or environmental impacts caused by leaking sewers (Amick and Burgess, 2000). A reason for the lack of knowledge is the still not confirmed correlation between exfiltration rate or volume and the impact on groundwater quality. A literature review on effects of exfiltration showed that information about the effect on groundwater is rare, site specific and very uncertain concerning the clear evidence of exfiltration from sewers as pollution source. Due to the uncertainty and limited data availability, simple approaches have been proposed. The methods are based on data on groundwater level, exfiltration rates and soil permeability (and soil type, respectively) as the most important factors for a potential groundwater impact. The exfiltration rate is calculated with the Leakage approach for all pipes where CCTV inspection data is available.

Four methods have been integrated in the assessment of the groundwater vulnerability: the American method DRASTIC (Cooper *et al.*, 1998), the method proposed by (Eaton and Zaporozec, 1997) and two methods that combine groundwater level and permeability on the one hand and exfiltration rate and groundwater level on the other hand.

In the USA the most widely used method is named DRASTIC, named for the seven factors considered in the method: depth to water, net recharge, aquifer media, soil media, topography, impact of vadose zone media, and hydraulic conductivity of the aquifer. With the scoring used in DRASTIC a simple map for the vulnerability can be produced.

The DRASTIC approach has 7 parameters. In CARE-S, similarly to the DRASTIC approach vulnerability is defined with 3 parameters as:

$$V = 5 \cdot G + 4 \cdot E + 2 \cdot S$$

Where:

V	=	Vulnerability of groundwater regarding to exfiltration
G	=	Rating for the Groundwater level beyond sewer level
E	=	Rating for Exfiltration rate
S	=	Rating for Soil type

Further factors as applied in the DRASTIC approach are not included into the assessment here to provide a simple procedure and to cope with the limited data availability.

The method by Eaton and Zaporozec uses Boolean logic operations to combine relevant parameters that influence the groundwater vulnerability. The parameters are the recharge rate, the depth to the groundwater table and the soil type. For the CARE-S software, the recharge rate has been replaced by the exfiltration rate to tailor the method to the needs of exfiltration assessment. Soil type, exfiltration rate and depth to the groundwater table are classified by the user prior to the application of the procedure, the values are then evaluated in a matrix, and the groundwater vulnerability will be assessed by using a classification matrix.

If data is not available for all three parameters the assessment is possible by using simple methods. The combination of exfiltration rate and groundwater level on one hand and the combination of groundwater level and permeability on the other hand will provide results for a simplified assessment. The calculation of the criteria is similar to the method by Eaton and Zaporozec.

Combined sewer overflow and impact on rivers

The assessment of receiving water impact is done in two steps. In the first step, which can be applied for water courses only, the sensitivity of the receiving water to wet weather impacts is assessed using a screening procedure proposed by the ATV (ATV Arbeitsgruppe 2.1.1, 1993). In this method the relation of number of inhabitants and mean flow in the receiving water indicates the pollution potential of the receiving water. The hydraulic impact is assessed by evaluating the relation between impervious areas of the catchment, and the hydrological catchment area of the river upstream from the urban catchment.

In the second step the impact-combined sewer overflows is estimated. The assessment is based on hydraulic data only. For load calculations a procedure described in the UPM manual (FWR, 1998) has been adopted to roughly calculate event mean concentration of overflows. Typical dry weather loads are assumed if no site specific data is available and a factor is used to calculated wet weather concentrations of pollutants. Furthermore the volume, frequency and duration are compared to user defined emission standards. For this comparison, standards to assess combined sewer overflows used in Europe are included in a software tool which is part of the CARE-S manager. Thus, the effects of rehabilitation strategies or the ageing of sewers on combined sewer overflows performance can be assessed in a relative way. Forecast on pollution or ambient water quality based criteria for the receiving waters cannot be done. The application of an assessment of surface waters is only possible if hydraulic simulations have been set up and run for long term scenarios in advance.

Tools

Two tools were developed during the activities of work-package 3.3. The intention was to consider the pollution of groundwater and surface water in the decision on rehabilitation strategies using CARE-S. The tools are integrated elements of the CARE-S rehabilitation manager and can also be used as stand alone applications. Each tool is delivered with a help file. Additionally the report "Tools in work-package 3.3" (Schulz, 2004) gives a summary of the theoretical background, explains input and output files, the acquisition of input data, the application of the different methods included in the tools and the use of outputs.

The software has been implemented in the CARE-S rehabilitation manager. Figure 3.8 shows the position of the WP 3.3 tools ("Groundwater assessment tool" and "Combined sewer overflow assessment tool") within the CARE-S procedure and the data exchange.

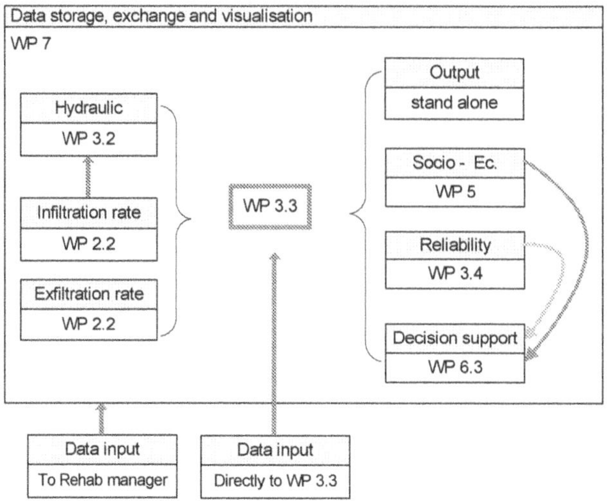

Figure 3.8 Work-package 3.3 in the CARE-S procedure

GAT (Groundwater Assessment Tool) is software to assess the vulnerability of groundwater concerning exfiltration from sewers. GAT is based on the rating methods described in the previous chapter and allows the identification of areas that are sensitive to groundwater pollution caused by exfiltration from sewers. Groundwater level data, soil type data and the results from the exfiltration model are input data to the procedure. The results are classified vulnerabilities on pipe level in ranges of low, moderate and high vulnerability or in numerical values and can be displayed in the CARE-S GIS window.

CAT (Combined sewer overflow Assessment Tool) is a software tool to assess the performance of combined sewer overflows. The tool compares results from hydraulic simulations to user defined standard values for a given number of years. Number, volume, load and duration of combined sewer overflows can be evaluated. The assessment is based on national legislation. Information on legislation in various European countries is included in the tool and can be updated by the user, other tools within CARE-S will use the outputs and the results can also be visualised as bar chart within CAT.

Additionally the tool allows the assessment of the potential impact the urban drainage system will have on watercourses due to wet weather discharges. The result is displayed to the user and will not be post-processed by other tools.

3.3.5 Example of integrated long term assessment

The main application of environmental criteria is the long term planning strategy rather than the selection of specific rehabilitation projects.

In an example study (Schulz *et al.*, 2004) the potential benefits of sewer pipe rehabilitation for the performance of the integrated drainage system was evaluated. The study focussed on the assessment of infiltration. The relation of sewer system status and the infiltration rate were assessed based on statistical analysis of 470 km of CCTV (Closed Circuit Television) inspected sewers of the city of Dresden. The analysis showed a correlation of infiltration rates and sewer condition of specific group of pipes. For these pipes an ageing model was run.

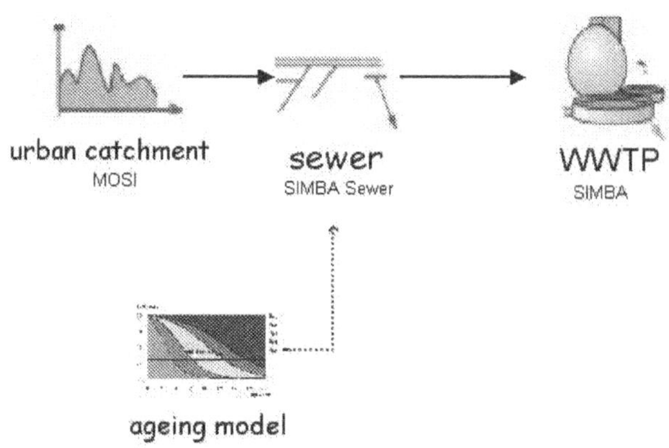

Figure 3.9 Integrated model

The potential reduction of infiltration rates and the consequent performance improvements of the urban wastewater system were simulated as a function of rehabilitation activities in the network in a hypothetical integrated model (Figure 3.9). In the model combined sewer overflow volume, frequency and duration were assessed. A similar procedure can be used for the evaluation of long-term rehabilitation strategies within CARE-S.

3.4 WORK – PACKAGE 3.4: COMBINING HYDRAULIC AND RELIABILITY MODELS

3.4.1 Task and reports

The service reliability at the system level reflects the probability that parts of the wastewater network are out of function because of accidents (clogging, pipe burst) sediment formation or because of hydraulic overload. A service reliability analysis includes a hydraulic, environmental and a reliability model, the latter predicting the probability of network capacity failure.

Task 3.4 has been focused in the evaluation of probability of occurrence of system failure events due to system deficiencies. The model developed within this task, aims to define and evaluate hydraulic and environmental criteria for the assessment of reliability aspects of the existing sewer system or for the examination of proposed urban drainage system rehabilitation scenarios. The assessment of environmental impacts is based on the national legislations.

This report summarises the methodology developed and the main results of this task, detailed study will be presented in Deliverable D10.

3.4.2 Hydraulic and Environmental Criteria

The WP3 first step to perform a rialiability analysis has been the definition of hydraulic and environmental criteria able to quantify somehow reliability parameters.

The criteria proposed and evaluated are classified as hydraulic and environmental. The list of hydraulic criteria includes:

- Frequency of filling a specific water level (A, B or C) if simulating single events; Probability of filling a specific water level (B, C or D) if simulating long term series;
- Link weighted capacity: $Q_{full\ j} / Q_{full\ Max}$
  Where :
  Q_{fullj} = flow at full section inside the j – pipe, flow in uniform steady condition inside the pipe. This value does not depend from the simulation done, but it is related with the total flow capacity of the pipe.
  $Q_{fullMax}$ = MAX($Q_{full\ j}$) from the comparison between all the Q_{full} evaluated for all the pipe, the maximum is selected.
- The ratio between these values is information of the weight of the specific pipe considered, in terms of flow capacity compared with all the other pipes. If the ratio is equal to 1, we are analysing the biggest pipe of the system and, of course, if flooding events are produced by this pipe the situation is really more critical than a flooding event produced by a pipe with a ratio value as 0,2, for example.
  We defined this criterion in order to weight in the simplest way the pipes responsibility to flooding events;
  Deviation from designed flow: $Q_{MAXevent\ j} / Q_{full\ j}$
  Where:
  $Q_{MAXevent\ j}$ is the maximum flow value inside the jth pipe during the event simulated;
  Q_{fullj} is flow at full section inside the j – pipe, flow in uniform steady condition inside the pipe.

  This ratio evaluates the possible insufficient flow capacity of a pipe, because the flooding event localized on a specific i-th - pipe could be produced by a downstream back-water or water jump due to an insufficient j-th pipe. When the ratio is greater than 1 the pipe is insufficient, in terms of flow capacity.
- Flow velocity comparison with standard in order to evaluate the link self cleansing process;
  While environmental criteria, derived from task 3.3 analysis, are:
  CSOs impacts criteria;
- Vulnerability of the groundwater;

In order to develop a methodology able to link both hydraulic and environmental indicators, the original work was based on classical statistical methods to determinate the occurrence of undesirable phenomena and hence possible risks, as presented in the following section.

3.4.3 Tool

The HELLMUD (**H**ydraulic and **E**nvironmental re**L**iabi**L**ity **M**odel of **U**rban **D**rainage) mathematical model is focused on service reliability, which reflects the probability of hydraulic efficiency and environmental impacts of a sewer system for one predetermined scenario. The model aims at a definition of several criteria that can be used for the assessment of reliability aspects on the existing sewer system or for the examination of proposed scenarios of the urban drainage system rehabilitation. As the piping represents the most critical system component with respect to hydraulic and environmental deficiencies, the criteria are highlighted at the pipe level (Table 3.5). The data are in relation to the risk assessment on combined sewer overflows, CSOs, which are perceived by the hydraulic model as commissioned marginal conditions of the mathematical modelling for each scenario. The assessment of environmental impacts is based on the national legislation, being defined in the project by CAT and GAT modules, task 3.3. The final table is saved in a text form "probability of deficiencies"(*.csv file) in the Project Manager database, being available to GIS and other uses. The criteria will be delivered via the CARE-S Rehabilitation Manager to other tools as input information for the decision-making process of sewer rehabilitation.

Table 3.5 Final HELLMUD report on the reliability of hydraulic deficiencies of the sewer network

Link ID	Hydraulic	Velocity	Infiltration	Exfiltration
248756	< 0,1 >	< 0,1 >	< 0,1 >	< 0,1 >
248757	< 0,1 >	< 0,1 >	< 0,1 >	< 0,1 >
248758	< 0,1 >	< 0,1 >	< 0,1 >	< 0,1 >
248759	< 0,1 >	< 0,1 >	< 0,1 >	< 0,1 >
248760	< 0,1 >	< 0,1 >	< 0,1 >	< 0,1 >
……..	< 0,1 >	< 0,1 >	< 0,1 >	< 0,1 >

Range (0 - links without any problem, 1 - links before the collapse)

The tool cannot be used without hydraulic simulation of the catchment. Environmental parameters for network probabilities of deficiency analyses are evaluated as defined by CAT and GAT tools. For a better understanding of the system conditions and status, final results will be visualized as maps. The CARE-S GIS system support is used to display the model results.

Format of input and output data for HELLMUD tool will be listed and described in Deliverable D10 which will be concluded by TU Brno in April 2005.

3.4.4 Hydraulic Performance

Results of hydraulic analysis are listed in column 1, table 3.5.

Thresholds for the hydraulic results cannot be defined without a previous collection of events and analysis of HELLMUD applications on a specific site. The pipe without hydraulic problem has hydraulic criterion "0". In all other cases, some hydraulic problems occur, and the closer is the value to "1", the worse failure is detected. The final decision concerning acceptable failure should make end user, e.g. according to amount of financial resources available. He can rehabilitate the worst links with evaluation near "1" and by steps continue to less value according to his financial possibilities. In advance, it is not possible to assess hard thresholds for any of four parameters listed on final report. HELLMUD compares the pipes in relation to each other and gives the comparison of all pipes in four parameters. It is not proper to synthesize the parameters into the only one parameter expressing total reliability of the link without sensitivity analysis of particular parameters. During the calculation, several (1-5) methods with result in the range of < 0,1 > are used and it is impossible to know in advance how many of them are used.

The hydraulic reliability strategy of the model distinguishes four classes (A, B, C, D) to classify all pipes (Figure 3.10). Class A includes design volume of the pipe, the whole range of the water levels inside the particular pipe, from the bottom to the top of the pipe (the behaviour of the sewer is reliable and safe, no overloading is detected). Class B covers safe storage volume of manholes; low overloading of the sewer is permitted. The class exceeds the top of the pipe and reaches to the critical level (see bellow). Class C spans between the critical level and ground surface. Dangerous volume is filled and the sewer is medium overloaded. It is supposed that in particular level between top of the pipe and surface level are situated basements of houses, the situation starts to be unsafe for the connected properties and their flooding via sanitary connections is expected. Finally, Class D involves all water levels exceeding the ground level; wastewater flows out from the sewer, floods the surface and can caused dangerous environmental problems. Classes B, C and D are considered to be unreliable.

The hydraulic vulnerability of the system is interpreted by means of so-called "critical level" (**h_{crit}**). This critical level is defined by the user (line between Class B and Class C, the level has got the range between the top of the pipe and the ground level of the elementary catchment of each pipe). The critical level is defined as a sewer network water level that, if exceeded, starts to bring losses on properties within the urbanized catchment in question. All the hydraulic events above the critical level are unsafe or dangerous and have to be controlled.

	Area	System behaviour		Status of the system
A	design	reliable within the range of design	safe	non-overloading
B	storage	unreliable	safe	low overloading
C	dangerous	unreliable	unsafe	medium overloading
D	flooding	unreliable	dangerous	high overloading

Figure 3.10 Pipe classification for assessing the vulnerability of hydraulic efficiency and environmental impacts of sewer systems.

The original endeavour of the authors in the development of a methodology for the assessment and linkage of hydraulic and environmental indicators consisted in the use of classical statistic methods for the determination of the occurrence of undesirable phenomena and hence possible risks. However, a necessary condition for the implementation of the procedure was the availability of a complete time series of the to-be-assessed parameters such as water levels, flow rates, velocities, spill frequency, volumes and pollution etc., for sufficiently long, representative period of time.

Similarly to many other cases of engineering practice where natural phenomena of accidental character both in time and space such as storms, waves, winds, floods, etc. are to be assessed, the probability of the exceeding of sought indicators was made with the use of the formulae below based on binomial probability distribution. Characteristics of binomial random variable are following:

(1) The experiment consists of *N* identical trials.
(2) There are only 2 possible outcomes on each trial ("event will occur" x "will not occur").
(3) The probability of "event will occur" is the same from trial to trial (*p*).
(4) The trials are independent.
(5) The binomial random variable *r* is the number of "event will occur" in *N* trials.

The binomial probability distribution:

$$P_{(N,r)} = \binom{N}{r} p^r (1-p)^{N-r} = P_{(N,r)} = \frac{N!}{r!(N-r)!} p^r (1-p)^{N-r} \qquad (8)$$

Where:

N = Number of trials (events) (In the case of water level classification in sewer, *N* is number of years, after which occurrence of given event is admitted). E.g. the useful design life of the structure or duration of insurance

r = Number of events in *N* trials. Total number of achieved water levels within classes A, B or C

$P(N,r)$ = Probability of *r* events occurring in *N* possible events

p = Probability of single event occurring in *N* possible events
= Return period = annual probability of exceedence.

If $\boldsymbol{p} = 1/\boldsymbol{T}$ (**T** – recurrence interval in years, see Table 2) and $\boldsymbol{r} = 0$, no floods will occur during the period *N* and the formula above is simplified into:

$$P_{(N,0)} = (1-p)^N \ ; \ p \in \langle 0.01 \ ; 1 \rangle \tag{9}$$

This is the probability that a flood (level A, B and C) with an annual exceedance probability will not be exceeded in the period of N years.

Generally, hydraulic risk R can be counted as complement-on-one to probability P. The result of hydraulic performance (Table 1, column 1) is hydraulic risk for each pipe assessed by means of probability that water level appropriate to the particular pipe is equal or higher then h_{crit}.

$R = 1 - P_{(N,0)}$, when $P_{(N,0)} = P(H \geq h_{crit})$

Where

R = Hydraulic risk for each pipe listed in Table 1, column 1
H = Water level appropriate to the particular pipe (link)
P, N, h_{crit} = See above

Table 3.6 shows percent occurrence of the risk of one or more exceedences during the design year period. The values of the risk are listed in % (interval of < 0,100 >) while in the Table 1 they are reduced to non-dimensional values in the range of < 0, 1 >. The basic formula and its parameters is explained and demonstrated by means of arrows. For example, a 20-year recurrence flood has a 10 percent chance of being exceeded within any 2-year period.

Table 3.6 Percent occurrence of the risk R [%] of one or more exceedences in a design year period

$$R = 1 - P_{(N,0)} = 1 - (1 - p)^N$$

Hydraulic reliability and the solution of annual probability of flood exceedence of particular system components is defined at two accuracy levels according to the user's approach to the hydrological data input into the validated deterministic simulation model (Figure 3.11):

- Historical rain data simulation - HRD (least **25** "worst" storms during 20 years)
- Single event simulations (optimum of 5; at least 3 events)

Preparation of hydraulic input data for HELLMUD is similar for HRD and SE simulations. The difference is amount of input data available. Firstly, dependence of relation $\boldsymbol{H/h_{dia}}$ on return period $\boldsymbol{p}$ (= annual exceedence probability) has to be found. $\boldsymbol{H}$ is water level appropriate to the particular link; $\boldsymbol{h_{dia}}$ is the diameter of the pipe. For input values (points in) of the simulation, mathematical formula for the dependence is found (the curve). Secondly, intersections of the curves with three given level classes (A, B and C) are calculated. The value of $\boldsymbol{H/h_{dia}}$ for class A is always "1", because the supposed water level for class A is top of the pipe, i.e. diameter. Class B corresponds to $\boldsymbol{h_{crit}}$ and class C corresponds to surface level. And finally, readings on x-axis are required results, return periods p(A), p(B) and p(C), which are listed in table as part of WP3 criteria results.

The SE simulation will be run on computation boundary data, which represents a return period of (for example) 5; 1; 0.2; 0.1 and 0.333 year^{-1}. For the single event simulation, it is accepted when the failure produced by the specific event has the same return period as the event.

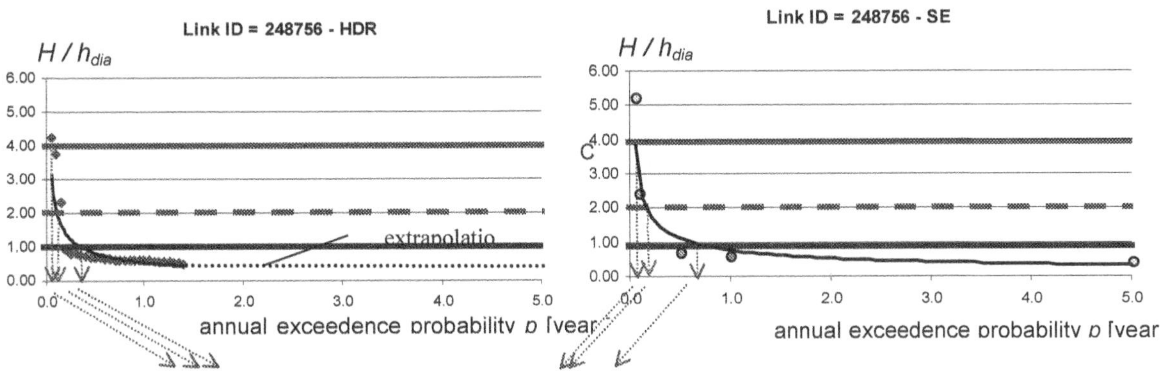

pipe	p(A)	p(B)	p(C)
248756	0.6 0.4	0.06	
248757	0.03 0.02	0.01	
24...			

A) the regression on historical rain data simulation (HRD)
B) the regression on single event simulation (SE), p = 5; 1; 0,2; 0,1 and 0,333 $year^{-1}$

Figure 3.11 Relationship between the water level (h / Pipe Diameter) and return period (link 248756)

The HRD simulation requires at least 25 "worst" storms during 20 years This means, that end user needs at least 20-years rain data series. According to Wussow, following formulae are recommended for assessment of storms and catastrophic storms (the shape of the boundary curve is drawn in):

catastrophic storms: $h = 2\sqrt{5t}$

storms: $h = \sqrt{5t}$

Where:

h [mm] = total precipitation depth
t [min] = duration of the rain

All rains are compared firstly with the first formula, and when within the 20-years rain data series is not sufficient number of catastrophic storms (25), the second formula is used for finding the rest of the heavy rains.

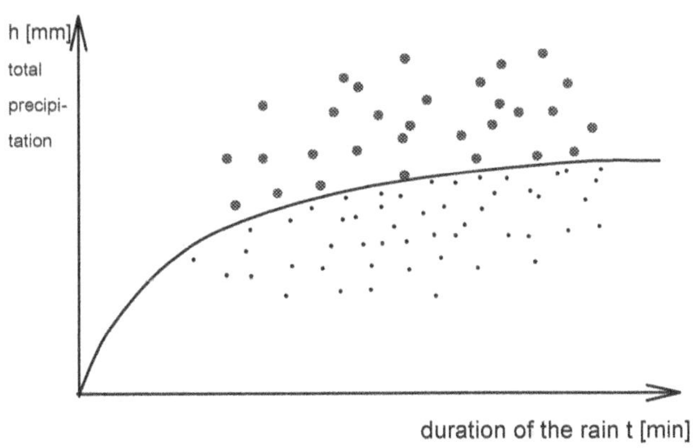

Figure 3.12 Assessment of the "worst" storms for historical rain data simulation

3.4.5 Velocity

Results of velocity analysis are listed in column 2, Table 3.4.

Two extreme situations concerning problems with velocity can occur in a drainage area. Firstly, too low velocity in sewer causes sedimentation of materials on the bottom and walls of the pipe, and consequently undesirable reduction of diameter. Secondly, high velocity in sewer will probably lead to abrasion of the pipe structure.

The HELLMUD module processes both low and high velocity problem and combines them, so that the result of the evaluation is deliverance whether or not the problem with velocity occurred. When problem with low velocity occurred, the value of pre-final evaluation is positive, within interval (0; 1>. In case of problem with too high velocity, the value of pre-final evaluation is negative, within interval < -1; 0). For pipe without any velocity problem the final value is "0". The final result for each link is absolute value of the number, i.e. in the range <0; 1>, where values near "1" indicated serious problems with velocity (but without specification "too low / too high").

There are no uniform and objective methods for assessment of minimum and maximum velocity (slope) in sewer. Many authors, standards and directions define different methods and HELLMUD module uses five of them. Following paragraphs describe general overview of the methods and principles included in the tool. End user can change the default value in a dialogue window.

Minimum Velocity Criteria

Assessment of minimum acceptable velocity in sewer is closely connected with assessment of minimum slope. Hydraulic, technical, operational and financial aspects should be taken into account. Too low slope in sewer heightens the diameter (the velocity is low too), the pipe is clogged by waste materials and has to be cleaned quite often. In the other hand, when the higher slope is designed, the pipe need not to be cleaned so often, but on a plain area wastewater pumping could be required which increases investments and operational costs.

HELLMUD module uses four methods for evaluation minimum velocity criteria:

1. Self-cleaning slope

Minimum self-cleaning slope depends on diameter according to recommended municipal standards based on operational experiences.

Table 3.7 Default minimal slope for each diameter in HELLMUD

DN [mm]	300	400	500	600	700	800	900	1000	1200	1400
Slope [‰]	4,90	4,63	4,43	4,27	4,15	4,03	3,89	3,85	3,00	3,00

If the slope of the link is lower then the minimal slope appropriate to diameter then the criterion is calculated by the following expression and SCS is of range (<0; 1>):

$$SCS = 1 - (\text{Links (i).Slope} / \text{MinimalVel.Slope(i)})$$

Where:

SCS	=	Self-cleaning slope criterion
Links (i).Slope	=	Slope of the i-th link
MinimalVel.Slope(i)	=	Min. slope which is appropriate to diameter of the i-th link.

Thresholds: If the full pipe velocity of the link is greater or equal to minimal slope then SCS = 0 (no problem occurred). When the slope is negative (Links (i).Slope < 0) then SCS = 1.

2. Shear Stress

Natural flush with certain shear stress (e.g. 4 Pa) occurring with certain annual return period (e.g. 5 year^{-1} for the combined sewer system) is able to regularly clean the sediments in sewer, which is sufficient for operation of the sewer. Shear stress τ_k depends on several parameters:

$$\tau_k = \rho \,.\, g \,.\, R \,.\, I \qquad \text{[Pa]}$$

Where:

ρ	=	Density of transported wastewater	[kg/m^3]
g	=	Gravitation constant	[m/s^2]
R	=	Hydraulic radius	[m]
I	=	Slope of the sewer	[-]

3. Frictional Velocity

According to English standard, minimal slope is the slope for which design velocity of wastewater is at least 0.75 m/s (variable parameter but it is highly recommended not to change the default value).

Criteria(i).frictVel = 1 - (SimulData(i).velDesign / MinimalVel.Frict)

Where:

Criteria(i).frictVel	=	Frictional velocity criterion
SimulData(i).velDesign	=	Design velocity
MinimalVel.Frict	=	Specified minimal frictional velocity (e.g. v = 0.75 m/s)

Thresholds:

Criteria(i).frictVel	=	0 (no problem)
Criteria(i).frictVel	=	1 (design velocity less then 0 m/s)

4. Full Pipe Velocity (Imhoff criterion)

According to German practice, minimal slope is the slope for which wastewater filling full pipe (full cross-section) has velocity 1 m/s (variable parameter but it is highly recommended not to change the default value).

If SimulData(i).Velcap < MinimalVel.FullPipe *Then*

Criteria(i).fullVel = 1 - (SimulData(i).Velcap / MinimalVel.FullPipe)

Where:

Criteria(i).fullVel	=	Full pipe velocity criterion
SimulData(i).Velcap	=	Capacity velocity (simulated velocity within full pipe)
MinimalVel.FullPipe	=	Specified minimal velocity of full pipe (e.g. v = 1 m/s)

Thresholds:

Criteria(i).fullVel	=	0 (no problem occurred)
Criteria(i).fullVel	=	1 (capacity velocity less then 0 m/s).

Maximum Velocity Criteria

Maximum velocity criteria are assessed according to particular pipe material (clay, asbestos cement, concrete, polyvinyl chlorine, polyethylene, iron, steel, stone, brick, unknown). Critical (default) value set in HELLMUD is 3m/s for clay, asbestos cement and concrete and 5 m/s for other materials. Sewer should be usually protected against abrasion, dynamic effects of wastewater and cavitation in case of higher velocity in the sewer.

End user can choose one or more criteria (or no one as well) for detection of problems with velocity. The final result is obtained by means of combination of chosen criteria. Table 3.8 shows results of comparison of some methods described above. While methods 1 and 3 individually have quite high evaluation error (26 and 18 %), their combination reduces the error to 8%.

Table 3.8 Comparison of selected methods for evaluation of velocity problems on the link *[Koudelk 2002]*

	1	2	3	
Criterion	Minimum self-cleaning slope	Critical Shear Stress	Minimum frictional velocity	Combination of criterion 1 and 3
	i_{min}	τ_k	v_t	$i_{min} + v_t$
Unit	[‰]	[Pa]	[m/s]	[‰ + m/s]
Criterion value	According to diameter	4.0	0.75	slope according to diameter + 0.75 m/s
Evaluation error	26 %	39 %	18 %	8 %

Infiltration

Results of hydraulic analysis are listed in column 3, Table 3.. This criterion uses loaded infiltration value per 1 m of pipe length and by means of simple calculation transforms it into weight – "contribution" of the link with regard to the whole network.

Final(i).infiltration = SimulData(i).qInfiltr * Links(i).Length / totInf

Where:

Final(i).infiltration	=	infiltration criterion
SimulData(i). qinfiltr	=	infiltration of i-th link (obtained from input data – file *simulation.csv)
Links(i).Length	=	length of the i-th link

totInf = total infiltration for the whole network (sum of all pipes infiltration)
Nlinks = total number of network links

Exfiltration

Results of hydraulic analysis are listed in column 4, Table 3..

In the frame of HELLMUD, exfiltration evaluation includes four vulnerability values (Eaton, Drastic, ExGround and PermGround) from GAT model in the range of real numbers of interval (0, 1). These vulnerability values are averaged according to number of used methods evaluating exfiltration within GAT model (max. 4 values). When no input data are available from GAT model, appropriate links are qualified as N/A.

CSOs

Data concerning CSOs are handled similarly as Exfiltration data. Input data are loaded from CAT model into HELLMUD via Rehab Manager.

From the HELLMUD point of view, CSOs, analysed via CAT tool, are included as boundary conditions within hydraulic model. It is presumed, that CSOs conditions are in compliance with national standards in term of the impact on water body and its pollution. HELLMUD checks distribution of storm water discharge in drainage area.

3.4.6 Conclusions

The inclusion of hydraulic and environmental criteria into the rehabilitation planning has been the main task of Workpackage 3. Within this task the focus was on the hydraulic behaviour of degraded pipes. Objectives of the assessment were surcharging of pipes in the catchment, combined sewer overflows and the receiving water quality during storms as well as overall reliability of the system. Innovation in WP3 has been obtained integrating new tools and procedures with the application of well-established hydraulic models. This approach has been used keeping into account end-users needs and, in particular, their will for not subverting the tools they use for analysing their systems. This choice greatly increased the project workload but it allowed for the development of tools that are suitable for application in the great majority of European cities.

REFERENCES

Amick R. S. and Burgess E. H. (2000). Exfiltration in sewer systems, National Risk Management Research Laboratory. US EPA.

ATV Arbeitsgruppe 2.1.1 (1993). Weitergehende Anforderungen an Mischwasserentlastungen. 1. Arbeitsbericht der ATV-Arbeitsgruppe 2.1.1. Korrespondenz Abwasser 40(5): 802 - 806.

Cooper B. S., Navulur K. C. S., Engel B. A. and Hahn L. (1998). http://pasture.ecn.purdue.edu/~aggrass/GROUNDWATER/drastic.html. 2004.

Eaton T. T. and Zaporozec A. (1997). Evaluation of groundwater vulnerability in an urbanizing area. Groundwater in the Urban Environment: Problems, Processes and Management. Chilton. Rotterdam, Balkema, 577-582.

FWR (1998). Urban Pollution Management Manual. Second Edition. Marlow, Foundation for Water Research.

Henze M., Gujer W., Mino T. and Loosdrecht M. C. M. (2000). Activated Sludge Models ASM1, ASM2, ASM2d and ASM3. IWA Scientific and Technical Report Nr. 9.

Schulz N. (2004). CARE-S Report: Tools in work package 3.3, TU Dresden.

Schulz N., Baur R. and Krebs P. (2004). Evaluation of groundwater vulnerability to support long term rehabilitation planning. Conference on Urban drainage modelling 09/15 - 09/15/2004, Dresden.

Schulz N. and Krebs P. (2004). CARE-S Deliverable D9: WP 3.3 - Environmental impacts of rehabilitation, TU Dresden.

4

Rehabilitation Technology information system

Carlos Montero, Angel Villanueva, Clabsa
Jaroslav Racklavsky, Vladimira Sulcova, Technical University of Brno

4.1 INTRODUCTION

The interest for rehabilitation in sewerage networks has increased in the last few decades, together with the developments of new techniques designed for this purpose. The reason is quite obvious: the modern sewerage and drainage networks of our cities that were constructed, in most cases by the end of 19th century and the beginning of 20th century, have aged unavoidably. The investment in rehabilitation, until recently, has been relatively poor, and the consequence of this is the deterioration of the system, the average age of which has increased to dangerously high values. This situation is not sustainable, and greater efforts should be made in the future in the rehabilitation of sewerage networks.

As it has been proved in many rehabilitation plans and studies, age is not the only factor that causes the failure of the system, and in many instances, not even the most critical factor. In fact, some elder conduits show better performance levels and structural state than newer pipes. The construction period is very stuck to the materials employed, their quality and the construction methods. It has been proved that some periods, that may be different from country to country, in which the main requirement is the construction of the infrastructure in as short time and at lowest cost possible, use bad construction techniques and the use of materials with shorter lifetime. These periods correspond, in many cases, to the post-war years. In the case of Barcelona, the so-called "Developing years" in the 60's, meant a large development of the sewerage network, but the use of concrete with a poor content of cement, has lead to short lifetime sewers that are today showing deficiencies. Similar stories happened in many other European countries in the years after 2 World War.

There are also some other noticeable factors that lead to the deterioration of the drainage and sewerage systems. As internal causes of sewer damage it can be can found:

- Quality of flow that may have low pH values (frequently originating in industrial spillage), that disaggregates concrete pipes;
- High quantity of flow and velocity that wastes the sewer invert;
- Hydraulic performance that may cause stress for internal pressure;
- Network age that presents natural wasting for use along years of low maintenance investment;
- Network construction time and sewer material, as it has been explained above;

- Constructive methods that carry associated trends presenting some particular defects, e.g. mining may lead to vault fissuring when soil over vault has not been well compacted and pipe works in pressure during rain events.

As external causes, it can be mentioned:
- Traffic overloads that may cause structural damages in shallow sewers;
- Soil overloads that may cause damages in deep and not properly designed sewers;
- Root intrusions that cause partial or total blockages, as well as fissuring or joint displacement;
- Mines, war shelters and tunnels, that change the soil stress state;
- Other buried services that may affect directly the sewers, piercing them partially, causing blockages and decreasing the hydraulic capacity;
- Type of surrounding soil that may be chemically aggressive to some pipe materials;
- Water table that may cause several problems in pipes under it; and
- Drinking water network leaking, which may wash out soil around pipes and also inflow in sewer network, causing different problems.

These causes of internal and external damages lead to diverse pathologies, mainly fissures, cracks, deformations, infiltrations, exfiltrations, obstacles, dislocation, abrasion, corrosion, pipe breakage and collapse.

The removal, or even the prevention of this damage, can be achieved by sewer rehabilitation.

4.1.1 Rehabilitation terminology

Rehabilitation is defined in the EN 752-1:1995 as *all the measures for restoring or upgrading the performance of existing drain and sewer systems*. In this standard, it is clearly stated that rehabilitation works include repair, renovation and replacement. It's important to emphasize this, because it is quite common to exclude repair from rehabilitation works, setting it as a maintenance task, together with other operational activities. It is also usual to use rehabilitation and renovation as synonymous, and confuse renovation with replacement. Figure 4.1 presents a brief summary of these concepts and their relationships. Definitions from standard are:
- Replacement: rehabilitation of an existing pipeline system by the installation of a new pipeline system, without incorporating the original fabric.
- Renovation: work incorporating all or part of the original fabric of the pipeline by means of which its current performance is improved.
- Repair: rectification of local damage.

Other usual works done in existing drain and sewer systems are the maintenance operations, addressed to *keep the pipeline system operational without the installation of additional fabric*. In this sense repair is in the border among rehabilitation and maintenance, as 'local rectification' may include or not additional fabric.

There are other concepts relating to rehabilitation (as renewal), substitution and reinforcement that can be looked up in the WRc Sewerage Rehabilitation Manual.

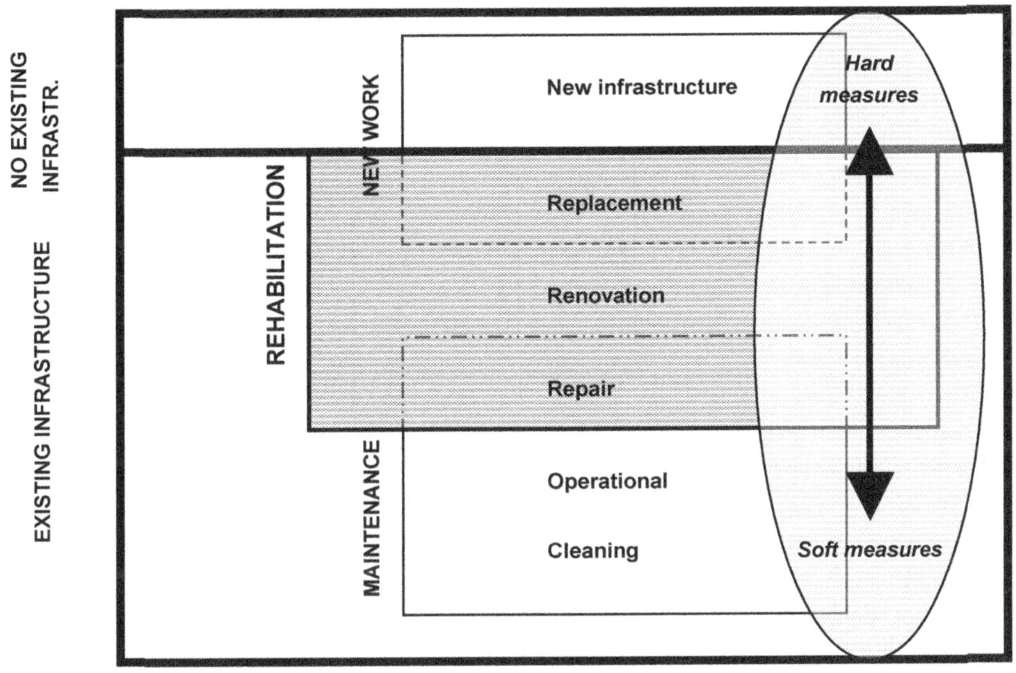

Figure 4.1. Terminology related to rehabilitation according to EN 752, and its relation with other operations in the drainage and sewerage systems

4.2 THE REHABILITATION TECHNOLOGIES DATABASE

The results of the rehabilitation technologies survey have produced over 50 different rehabilitation technologies, and for each one 30 characteristics have been defined.

The target of the CARE-S project is to help to rehabilitate 'the right sewer, at the right time, with the right technology'. This requires, among other large amounts of information and calculations, access to the available techniques and their main characteristics. This information must be available in a dynamic way, that is, data must be stored in an easy way to query files, organize and arrange. This is an important issue, because the aim of this surveying task is not just to establish a list of the state-of-the-art of rehabilitation technologies, but provide a useful tool to other tasks developed in the CARE-S project.

The information was collated in a big chart, in a first instance, and later exported to a database. A database file makes possible a dynamic inquiry from the Multi Criteria Decision Tool created in the CARE-S project, providing the list of techniques able to solve a specific defect in the sewer or drainage system (under defined environment conditions and dimensions - see *Applicability conditions or requirements*). This database also provides data for the calculation of socio-economic costs, like the environmental costs (see *Environmental impacts*), the disturbance to citizens (working area required, time to develop the works, etc.) and others (see *Technique rates and characteristics*).

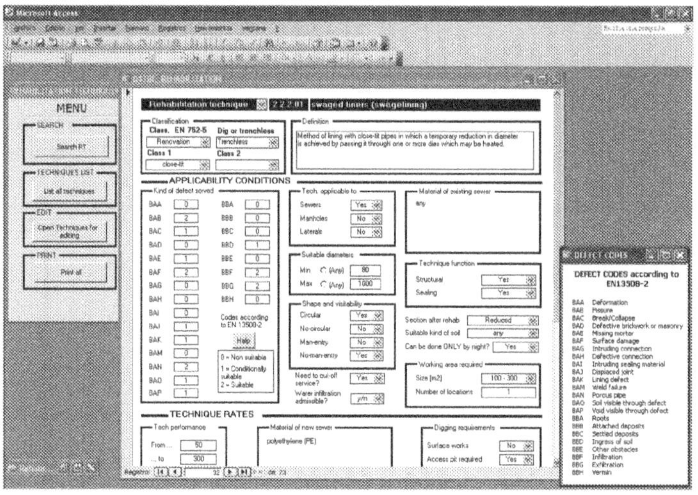

Figure 4.2 View of the Rehabilitation technologies database. From left to right, the menu box, partial view of a rehabilitation technique card and the helping defect codes list box

The software used is Microsoft® Access, due to its availability (it's the most common database software, attainable from any end-user), its compatibility with other tools of CARE-S project and the possibility of providing this database as a stand-alone tool. The database can be updated with new techniques or characteristics easily, giving the end-user the possibility to adjust this tool according to their experience and local requirements.

The Rehabilitation Technologies database is structured following the lines explained in 'Rehabilitation classification'. It includes a brief definition of each technique, a space for comments, the possibility to add commercial names of installers linked with the technologies, a help menu with the list of the defect codes as well as a pop-up menu with this information. A menu box provides some facilities for listing, editing and printing the techniques.

This tool has been improved with a search engine. Searches can be done on the most significant parameters:

- Kind of technique: rehabilitation or operational.
- Dig or trenchless.
- Repair, renovation or replacement.
- Kind of defect solved by the technology, according to Standard coding (with the help button to remind the codes meaning)
- Sealing and/or structural technique
- Suitable diameter: minimum and maximum, with logical operators (<, =, >)
- Technique for sewers, manholes and/or lateral connections.
- Results are provided in a list, and technologies can be acceded directly by double clicking.

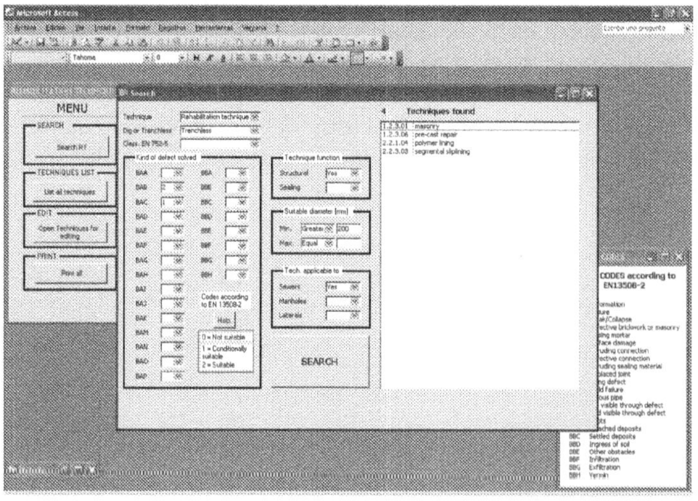

Figure 4.3 Search engine of the Rehabilitation techniques database.

4.2.1 Rehabilitation classification in the database

The first step was to survey all of the rehabilitation technologies available in the market, which were classified in repair, renovation or replacement. Technologies were also classified and characterized systematically further in more detail. A main sub-division between dig (or trench) and trenchless technologies was made. Over 50 rehabilitation techniques were identified.

To make homogeneous comparison easier and characterization of the different rehabilitation technologies, many characteristics were searched and assessed methodically for each technique. These characteristics, contained in the rehabilitation technologies 'specifications card' (entries of the database), were divided into three groups, according to the information contained:

Applicability conditions or requirements

(Characteristics that define a rehabilitation technique as suitable, or not suitable, to solve a rehabilitation problem.)

These characteristics are used by Multicriteria Decision System to offer the end-user only the rehabilitation technologies that are suitable to solve the particular problem, before ranking them. Some of these basic characteristics are:

Kind of defect solved: Obviously, one of the most important characteristics of a technique in order to decide about its suitability to solve a sewer defect appropriately. To achieve this purpose, possible defects in sewers have been classified according to standard EN-13508-2, Condition of drain & sewer systems outside buildings, Visual Inspection coding system.

The rehabilitation technologies (RT) are referred to all these defects according if they are suitable (index "2"), conditionally suitable (index "1") or not suitable (index "0"). When a RT is indexed as suitable it means that the technique is appropriate to solve the defect, and it's usually employed to that aim, i.e., RT is adequate and reasonable to solve the defect in a cost-benefit context. A RT is considered as conditionally suitable when it may be able to solve the problem under specific conditions or it's reasonable in certain cases.

Shape characteristic informs about the suitability of the RT to be applied to different parts of the asset. Manholes and lateral connections (from houses or gullies) are also parts of the network that require rehabilitation, even that in general these does not suffer the same loading conditions that the sewer itself. Most of the techniques are suitable for sewers, but a few of them only are used in small diameters, and so these are considered to be only useful for lateral connections.

Original material: Many rehabilitation techniques are only suitable for specific sewer materials, as fragile (or at least not-ductile), and consequently some techniques will be discarded if the existing sewer material is not listed.

Size of existing sewer, indicates the minimum and maximum size the technique can cope with. These measures may vary in local markets, as not all the sizes (diameters) are available in all the regions. The measures exposed are intended to be the extreme values of the European market. When marked as "min" or "max", it means that the applicability of the technique is independent of the minimum/maximum size of the sewer. The minimum size for a 'man-entry sewer'has been establish at 1600 mm, and consequently all the man-made repairs mark this size as minimum.

Sealing or structural technique. Most, but not all, rehabilitations technologies are designed to be watertight, and this is one of the main characteristics of any flow conduit, that avoids infiltration and exfiltration. The other major characteristic of a conduit is that it must be able to support the design loads. Techniques are characterized as if they are able to increase significantly the structural strength of the conduit, and also if the technique avoids infiltration, exfiltration, exudation and other undesirable phenomena of fluid transference.

Size after rehabilitation: One of the reasons that lead to the rehabilitation of a sewer may be poor hydraulic capacity. In other occasions, the hydraulic performance needs to be improved due to different reasons. That's because it's important to know the change in the cross section of the sewer after rehabilitation. This characteristic has been characterized as "not changed", when the rehabilitation works does not influence in the sewer internal size; "increased" if the section is significantly enlarged its diameter; or "reduced" for the rehabilitation techniques that usually have as consequence a reduction of sewer cross section. In this sense, it must be remarked that due to the large variety of installers, materials and conditions even for the same technique, it's almost impossible to assess an exact value for this characteristic. Different manufacturers present also significant variations for similar conditions. As consequence of this, the simplification made grouping the changes in simple categories is convenient. Some techniques may be marked as combinations of these factors.

Need to cut off temporally the service connection: if the technique requires to stop the flow of the lateral connections to be carried out. This interruption of service may be sometimes impossible, others inconvenient, so that this characteristic may become sometimes a decisive factor for the sewerage manager.

Application under groundwater level or leakage admissible: this field provides the information of the requirements of watertightness of the RT under two different scopes: whether the technique is applicable under the watertable or significant infiltration is admissible during rehabilitation works. This aspect is more addressed for no-dig technologies.

Suitable kind of soil: soil surrounding does not compromise the suitability of most rehabilitation techniques, but in few cases, techniques should be carried out in favourable soil conditions. This can be different for diverse techniques, and so drilling technologies require cohesive soils, but present very low performance in rocky soils. This does not intend to force sewerage manager to invest in large geological data collection campaigns in a master plan phase. More simple information is only requested, and experience and knowledge of the territory by managers is usually enough for the purposes of this selection. The soils have been divided in a very simple way, depending on its hardness: soft, medium, hard, any and combination of these factors.

Working area required: Urban areas in general, and city centres in particular, are in European countries very busy and quite often it is difficult to find space to carry out the works. Rural areas do not have this problem. This parameter pretends to point an idea of the room required for the work, under usual conditions and rehabilitated lengths. The information of the area required is completed with the number of locations in which this area is divided. Some techniques allow or require distribute the working area in several sites, usually two manholes or access pits. The categories are: Small, for occupied area less than 100m2 Medium, for areas between 100 and 300m2 Large, for areas between 300 and 800m2 Whole length, for occupied space over 800m2 and for trench technologies.

Technique rates and characteristics

This group contains technology characteristics that are not usually decisive, but give useful additional information, However, these data give very useful information for ranking the techniques, as gives parameters that allow calculation of important data, as for example an estimation of work duration, the requirement of auxiliary works, etc.

Maximum length from each location. This parameter provides an estimation of the maximum length of sewer that each technology is able to rehabilitate from one single working location (usually, a manhole or an access pit). This gives an idea of the number of working locations required in works of special length. Value is given in meters, and measured only in one direction from the location, i.e., downstream or upstream. As rehabilitation techniques works are generally independent of the sense (direction), in most cases the length of pipe able to be rehabilitated from one location is the double of the "maximum length from each location" given value. Results presented are estimation, and different manufacturers may provide different values. Even more, the same rehabilitation technique may vary, and usually do, the working length depending on the pipe diameter: bigger diameter, shorter length. This is marked in the chart with a value in parenthesis. Some repair techniques affect very short lengths, and in consequence a minimum length of one meter has been imputed. On the other side, rehabilitation methods with virtually no technical limitation, as for example trenching methods or man-made repairs, have been catalogued as unlimited, that should be understood only in that sense.

Performance of the technology (rate of length constructed per day). The performance of the rehabilitation technologies is the average length of rehabilitated conduit per day. This value has been estimated from diverse sources, and includes the standard auxiliary works required by rehabilitation. This characteristic does not include the time required for any special operation before or after the works themselves, The performance has been calculated as an average, consequently, particular works, or installers, may increase or decrease significantly this time depending on their skill or technical improvements to be more competitive, but strongest variations may come from factors external to the rehabilitation technology, like working time restrictions due to traffic cut off, breakdowns, pace of materials supply and a wide variety of unexpected problems.

Material of the rehabilitated sewer. The material of the new sewer is of significant importance in order to know the new roughness of the sewer inside, which is an important parameter for the calculation of the sewer hydraulic capacity.

Digging requirements (pavement demolition, access pit or trench required). It's a figure that expresses the excavation activities during the works, described by means of three parameters: "surface works" that include any kind of pavement demolition, excavation, etc. that affect the surface ground; in these cases, the technique operates usually from the manholes (service entries). "Access pit" for techniques that require the digging of a restrained size hole in the ground, usual for pipejacking and sliplinning techniques. "Dig or trench" are rehabilitation technologies which works involve significant digging (pits excluded) or trenching.

Need of digging for reconnecting laterals, is a significant data, because some trenchless technologies force (no other option) to ground excavation, loosing part of no-dig technologies advantages.

Requires later reopening of laterals. Many techniques block the lateral connections. These must be re-opened latter with a robot in no-man entry sewers or manually in bigger conduits. This makes extra charges and time consumption, that must be considered for direct and indirect cost calculation.

Night technologies. Techniques classified as "night technology" when is possible to develop the works in night periods, remaining the surface with no significant disturbances during the day. This characteristic is an important issue in order to assess the social cost of rehabilitation works. The reason is obvious: during the night the traffic disturbance caused by lanes closing, streets closed, etc. is lower, and so it is the indirect costs.

Environmental impact: environmental characteristics are assessed providing additional information for indirect costs calculation. Next characteristics are measured in a very basic scale, and it's not necessary to say that different local conditions, manufacturers and installers may cause greater or lower impacts.

Environmental impact of works and materials, shows a basic and general assessment of the environmental impact caused directly by the works, including the materials employed.

Impact on groundwater quality, there are techniques (grouting, injection of chemicals) that have been rejected in specific conditions due to its pollution potential, that can cause contamination of the watertable or soil close to the sewer. It becomes essential to know if the rehabilitation technology is potentially dangerous for the groundwater when this water is used as a resource.

Structural impact on the surrounding buildings or networks. In densely built areas, as most of European city centers are, the underground is occupied by many urban services, and most times urban networks run in parallel and cross each other along the streets. Digging a trench or using a soil aggressive technique (techniques that make heavy vibrations in the ground and/or compact the soil) can damage the infrastructures nearby, including the foundations of the buildings.

Acoustic pollution (noise), is an important issue for social cost assessment, and combined with "night technologies" may rule out rehabilitation technologies that could be done by night, but are not used because of the noise.

Air pollution, thought as dust produced by earth works, not as pollution due to exhaust of machinery. This dust causes citizens complaints.

All this characteristics given for each rehabilitation technique provide plenty of information that allows the end-user to compare the benefits and drawbacks of each technique, knocking out the not suitable ones for specific problems.

4.2.2 State-of-the-art in rehabilitation technologies: List of techniques

It has been pointed before that the rehabilitation technologies are a rising market, and new techniques and improvement in existing ones appear each year. There is a large list of rehabilitation technologies available in the market, able to solve a wide range of defects. Trenchless technologies are sometimes the preferred option, due to its lower social cost, particularly in densely urbanized areas. The rehabilitation technologies database clusters them in next categories:

Trench technologies. The most common and widely used technologies for rehabilitation works (including repair, renovation and replacement):

- *Open cut*: Trenching is simple in most places, does not require skill personnel and its cost is moderate, working in proper conditions.
- *Semi-open cut*: includes shoring, trench box and sheet pilling. Higher cost, but safer working conditions.

Repair techniques. Present next trenchless technologies:

- *Robotic repairs*: the main advantage is that robots operate in small conduits. They are each day more skilled and able to repair more laborious works. Their main drawback is, still, their price, due to the high level technology they use and the trained personnel robots require for operate. Most common are the root cutting and the milling robot, but also there are injection, pointing and rerounding robots.
- *Short-liners*: installed with packers in no-man entry conduits, these techniques have found a nice niche in the market for repairing local damages like defective joints and very localized watertightness problems.

- *Man-made repairs:* widely used in man-entry sewers (usually, higher than 1.5-1.6m). Allow a wide variety of techniques, at an affordable cost. Mort common man-made repairs are masonry, pointing, injection and invert repair (in situ or precast).
- *Grouting:* trenchless technologies consisting in injecting chemicals, resins or mortar addressed to repair cracks and fractures.

Renovation techniques. The no-dig representatives are:

- *Coat lining*: the idea is to project or spray a material that creates a film that sticks on conduit inside surface. Projected material can be cement mortar, sometimes reinforced, and/or polymers.
- *Sliplining:* it's a quite extended technique, with the simple idea of introducing a new pipe inside the old one. This can be done with short pipes (small enough to be introduced in the pipeline through the manholes), long pipes or continuous pipes. The main drawback is that the new pipe is significantly smaller than the old one. The gap between the old pipe and the new one (ring gap) can be filled with mortar slurry.
- *Close-fit lining*: pre-deformed plastic pipes, with reduced external dimension that allows them to fit in the existing pipe. Once inside the old pipe, the folded pipe restores its cross section, generally by heating. In this category it could be placed swageling and folded liners (fold&form lining). These methods reduce the ring gap that is common in the sliplining techniques.
- *Spirally wound lining*: the new pipe is a PVC continuous helix, constructed inside the old one.
- *Cured in place pipe (CIPP) or soft reversion lining*: one of the most extended renovation techniques in which the biggest difference with the other lining techniques is that the new material is a soft liner impregnated with a resin. The soft liner is introduced in the sewer easily (usually through a manhole) and then blown up to contact with the inside walls of the old pipe. The resin is then cured, using hot water, steam or UV light.

Replacement techniques. Also set out a list of trenchless solutions:

- *On-line replacement*, destructing the old pipe: the king on this category is probably the pipebursting (pipecracking) technique, consisting in introducing a breaking cone (bursting head) in the old sewer, with a plastic pipe attached in its tail. In this way, the old pipe is being destructed at the same time that the new pipe is installed. His can be done only in non-ductile existing conduits. A variation for ductile materials is the pipe splitting technique, which has a cutting blade in the cone-head. If an inverted cone is used, the material gets inside the new sewer and can be remove from soil. This variation is called pipe crushing or pipe implosion.
- Some techniques that can be also set in this same category are spin-off tunnelling, with noticeable modifications and prepared for small diameters, as pipe eating, pipe reaming and the "controlled line and grade system"(CLG).
- Other technologies come from pipejacking, like pipe ejection (also called modified pipe jacking), pipe extraction (or modified static pull), and pipe pulling.
- *Non-steered off-line replacement*: these techniques are used off-line the existing pipeline, this means that are also used for other proposes that rehabilitation. The techniques are relatively simple, and the installer can't guide the direction of the new pipe, that is supposed to make a straight path. Some examples are: rod pushing, impact moling, steerable (partially) moling, pipe ramming (or impact ramming) and auger boring.
- *Steered off-line replacement*: these techniques present the important advantage that the new pipe that is being installed can be steered during the works are taking place. This steering is limited, and it's accuracy depends on technique limitations, the operator skill and other external factors, like the lake of homogeneity of the soil. Some categories could be differenced.
- *Directional drilling*: are the most affordable techniques under this category, used for small diameters. It can be mentioned as examples: guided auger boring, horizontal directional drilling (HDD) and guided drilling (mini-HDD). HDD techniques are quite extended, but are not suitable for gravity systems, because the installed pipe profile is parabolic; but it's widely used in pressurized systems and telecommunications pipes (carrying wires), and has proved to be quite useful to cross buried obstacles (if already knew).
- *Pipe jacking*: used for bigger diameters, there are some variations of conventional pipe jacking, like low load pipe jacking and thrust jacking. The idea is quite simple: push a pipe into the soil from a pit, with the help of hydraulic machinery, and remove the soil once the pipe, or a length of it, has been installed.

- *Tunnelling*: depending on required diameter, it can be used microtunelling, minitunelling, tunnelling or shields, for very large diameters. The last two techniques are rare in drainage networks; they are used for transportation infrastructures. The tunnelling heads are in constant development, although they are quite use-specific: the cutting front must be adapted to the kind of soil that is supposed to be found to achieve a god performance (ratio meters/day). The cost of tunnelling is high, and its use in drainage systems is reduced to long new trunk sewers under cities with highly occupied surface.

Figure 4.4 Replacement trenchless techniques. Tunnelling (mining) on the left, pipe jacking on the right.

4.3 OPERATIONAL METHODS

4.3.1 Introduction

Most common problems found in the sewerage systems are related to maintenance issues, sediments, attached deposits, roots, etc. Rehabilitation techniques are clearly not cost efficient to solve most of them. This task has collected the most usual operational methods used in Europe to clean sewers.

There are several processes of cleaning sewer systems and sewer connections. Generally, they can be divided into hydraulic, mechanical, combined, chemical and biological processes. Such methods that do not require the presence of operatives inside the sewer are generally recommended. When cleaning the sewers by means of high-pressure jetting vehicles, water pressure that does not damage the structure of the sewer (sometimes the sewer is damaged when water jet is used, this is mainly the case of plastic and pitch fibre sewers) should be used. Mechanical tools are recommended where it is necessary to remove hard clogging objects in non man-entry sewers. If incrustation and solid sediments are present in the sewer they should be removed by a milling cutter, a chisel or a pneumatically driven pipe knife. It is often necessary to remove protuding pipes and jammed objects projecting into the sewer by means of a robot that is remotely controlled via a TV camera. A similar tool can be used to remove roots growing through the sewer. However, this method is not especially efficient as, once cut, the roots quickly re-grow again and in some cases the trees wither after having the roots cut off. It is more suitable to use a biological-septic method of root removal that causes the roots to rot away in the immediate vicinity of the sewer without damage to the tree.

Table 4.1 Division of detritus in the sewer systems

A – sediments	Mineral particles (sand, stones, etc.) Organic origin (rest of foodstuff, plastic materials, paper etc.) Other substances (tins, pieces of glass, etc.)
B – detritus adhering to the surface	Mechanical (clayey and earthy particles, paper, etc.) Chemical (grease and oil, chemical detergents) Encrustation, scale
C – solids particles projecting over the designed profile	Due to poor workmanship (mortar at joints projecting over the sealing, displaced pipe ends) Due to additional work (protruding connections, other underground services crossing the profile) Tree roots
D – pollution soaked into the pipe material	Non-reacting with the material (water, etc.) Chemical reacting (acids, alkalis, etc.)

Factors decisive for the selection of the cleaning process are:

- accessibility to the sewer,
- depth of soil cover over the pipeline,
- type of cross section and dimensions of the sewer,
- distance between the access points (manholes),
- changes in the diameter or misalignment /relocating,
- type of sewer material,
- structural condition,
- weather condition (rain, snow, frost),
- traffic situation,
- service conditions (blockage, sediments, etc.).

4.3.2 Occurrence of pollution according to the type of sewer

The selected cleaning process must correspond to the composition of sediments, also depending on the type of sewer.

During rain events, water flows at a high speed in storm water sewer. This speed easily released sediments that are subsequently transported.

As regards foul sewers, the discharge conditions are almost uniform, which results in continuous settling in dependence on the speed of the water flow. The speed of water flowing in the pipe changes with the rising volume of sediments, viscous resistance increases based on the pipe geometry, which reduces the sediment-transport capacity. This results in accumulation of sediments in the sewer with consequent reduction in the sediment-transport capacity of the sewer. These sediments must be removed.

Under special circumstances the sediments may harden (above all, in industrial sewerage systems), which makes it considerably more difficult to remove these sediments (such as limescales sediments, encrustation, formation of crystals, etc).

4.3.3 Extraction of material from the pipelines

If the flow rate rises, it is not possible to fully extract all material transported to the manholes, which results in pollution of the next section of the pipeline. For this reason, the sewer system should always be cleaned from the upstream end down towards the wastewater treatment plant. The quality and velocity of cleaning depends on the power of pressurised water. As the pressure of flushing water must be limited to avoid damaging the pipe, higher efficiency may be achieved by increasing the volume of flushing water.

The cleaning frequency depends on the pipe gradient i.e. on the settleability of sediments in the pipeline.

4.3.4 Basic technology of sewer cleaning

Operational methods have been clustered into the following groups:

- Rinsing / flushing
- High-pressure water jetting
- Mechanical processes
- Other processes
- Combined processes

4.3.5 Flushing Process

This is the oldest sewer cleaning system. Based on the flushing process, this cleaning system is divided into two methods. Both these methods are based on removal of sediments using increased velocity in the pipeline.

- *Surge Flushing:* the principle of the process is based on retaining water (wastewater or drinking water) in the shafts or flushing tanks. Once the required level in the shaft is achieved, a 'clack' valve is opened and the sewer section is flushed. The flushing effect depends on the level and volume of retained water and the pipe gradient.
- *Back-up flushing*: special tools are inserted into the sewer system (flushing shields, flushing boats, cleaning balls etc.), which create plugs in the stream retaining (causing backwater) the wastewater. Retained water presses on the equipment; the equipment moves and pushes sediments ahead.

4.3.6 High-Pressure Water Jetting Process

This is currently the most widely used method of removing sediments (hardened and settled sediments) from sewer systems. However, this process is not usually used for removing obstacles from the stream e.g. connections, artificial obstacles, roots etc. Mechanical cleaning methods must be used in such cases.

- *High-pressure water jetting:* the flushing water is conveyed, through high-pressure pumps, from a tank into a nozzle with jets that generate high water pressure acting on the pipe walls. The hydraulic effect of the jet pressure (pressure up to 10 - 34 MPa) cleans the pipes by means of disintegrating and whirling of sediments. In cases where the jet pressure is not sufficient and it is necessary to remove very hard sediments and in-grown roots, special pressure nozzles, milling adjusting pieces and heads or chain cleaners are used. At the first cleaning stage, the hydrodynamic effect pushes the cleaning head from the starting access point or manhole upstream into the target manhole. At the second stage, the cleaning head is slowly pulled downstream from the target manhole to the starting manhole. The hydraulic effect of the jet moves the sediments gradually towards the downstream manhole from where the sediments are sucked into a cleaning vehicle.

  The number of jets depends on the inner diameter of the pipe, the type of pollution and the volume of cleaning water. Currently, the high-pressure cleaning equipment is mounted on vehicles.

  Unprofessional application of high-pressure water cleaning may result in damage to the pipe in the form of grooves, cracks and holes in the pipe wall or the lining. When designing the operating parameters of the cleaning process it is necessary to consider pipe material, wall thickness and condition of the pipe.
- *Sewer cleaning by suction:* in large diameter man-entry sewers it can be effective to suck the material out of the pipes, assuming certain conditions of volume of water, type and amount of sediments. In order to do so, the operatives must be inside the pipe. This cleaning process requires a minimum water flow rate (it must be re-pumped). Big pieces of sediment or obstruction must be removed manually. The disadvantages of this cleaning process consist of the aforementioned restriction of the water flow rate and relatively costly safety measures.
- *Kanal-Jumbo:* this process eliminates the disadvantages of sucking because a high-performance, hydraulically driven pump is installed right in the pipeline. A stone crusher and cloth ripper are installed before the pump. Nozzles and jets are directed towards the bottom of the pipe, according to the profile, in order to whirl the sediments and create suspension. The entire unit is mounted on a circular gear and is pulled by a rope into the pipe from where it transports a mixture of water and released impurities to the

surface via a suction hose. On the surface, the extracted material is separated from the water stream using hydrocyclones and, following further treatment, the water is re-used for cleaning purposes.

4.3.7 Mechanical cleaning

This is one of the principal tasks of the sewer network operators in order to guarantee full design capacity of the pipe systems is maintained and full functioning of the water-related structures is preserved. Mechanical cleaning is used to disintegrate and remove sediments, based on the type of sediments, incrustation, availability of suitable machines and the distance between auxiliary entries.

- *Manual cleaning using auxiliary tools:* it is limited by the size, skills, experience and equipment of the engineers as well as by the relevant health and safety regulations. This work concerns cleaning of man-entry profiles from inwash and sediments. The working activities include disintegrating, loading and transporting of these sediments. Tools include pick axes, pick hammers, blasting, tension-suspended scrapers and loading machines. Incrustation-crystallised precipitates can also be removed manually, by grinding or sand blasting using high-pressure aggregates, from man-entry sewers based on the type of the pipe material and refurbishment method.
- *Cleaning using standard cleaning equipment:* this consists of several working stages and relevant equipment. Firstly, the sediments and inwash is disintegrated and expanded. Secondly, this material is removed from the pipe profile, i.e. the sediments and inwash are transported into or towards a shaft. The equipment that pulls can be diverse: buckets, chains, sewer ploughs, scrapers (mainly for incrustation), go-devils made of foamed materials, sleeve and sewer anchor, sewer bucket machine and loading machines, canal diggers, sewer scrapers, canal cleaning brushes, rubber circular brushes- arms. The equipment that drills is provided with metal spirals.
- *Cleaning using special equipment:* in recent years, equipment achieving high level of cleaning, above all in small profiles have been developed and is used to remove hardened sediments projecting over connection pipes and artificial obstacles and roots. Based on the principle, they can be divided into following equipment:
    - Impact equipment – blow. These include pipe core drills for circular pipes and egg-shaped profiles. Compressed air is used to drive the equipment and released sediments are carried in pressurized water and extracted from the shaft. Some equipment is fixed to the central rotary head by chains that remove the sediments by impacting. The drive of the equipment is pneumatic or hydraulic.
    - Drilling and milling equipment – rotative. This equipment is remote controlled and monitored by TV cameras. It is used to remove solid sediments and projecting outlets of connection branches and other obstacles in the profiles as well as roots. The drive is hydraulic.
    - Drilling and milling equipment – rotative (blow). This category of equipment includes automatic blow drilling jets used to remove solid sediments. The jets ensure movement to removal of the sediments, cleaning of the inner pipe or canal wall and transport of disintegrated sediments. The rotary and blow movement of the drilling equipment is driven by high-pressure water towards the front jets. The drilling equipment produces impacts on the sediments and incrustation based on the working pressure and additional rotation acts like a milling machine. The pressurized water used for the equipment is then used to flush the drill blade. Cutting knives fitted to a steel drilling head are used as the drilling tool.
- Cutting equipment – mechanical: cutting equipment is used to level projecting outlets of connections and other artificial obstacles such as roots, observed by a TV camera. Mechanical equipment includes the pipe saw robot. It is pneumatically fixed in the working place. Mechanical robots offer their own drives and various and multiple operations, root shears, clamping tongs, pneumatic hammer, saws made of hard metal, spring knife for removing roots, etc. The equipment is remote controlled.
- Cutting equipment – high-pressure water jet: cutting equipment uses a high-pressure water jet to cut all obstacles in a continuous stream and removes more extensive incrustation where mechanical removing is not sufficient. The work is inspected by a TV camera. The high-pressure jet is located on a movable arm and a movable gear.
- Sand blasting equipment: in order to achieve and create an especially high degree of perfect surface, mainly in metal pipes (steel cast iron), surface sand blasting is used.

4.3.8 Other operational methods

These are based on the principle of chemical reactions or biological methods.

- *Increasing the flow rate by adding air or polymer*: this method is used to clean inverted siphons, other narrow spaces and in pressure sewer systems. Added air generates higher speed of the medium in the pipe and as a result the drifting speed increase. This is conditioned by creation of 'pockets' in the pipe axis, where the air creates minute droplets from closed water, resulting in a water cylinder facilitating mixing of water and sediments. Another possibility is increasing the flow rate is the addition of polymers. The benefit of the polymer means reduction of energy looses (impact on flow turbulence). This increases the velocity changing viscosity and also reducing friction on the pipe surface. Increased drifting velocity is an advantage for the transport of solid particles. Of course, it is necessary to assess the effect of polymers on water. Therefore they are
- *Chemical cleaning methods.* These chemical methods are limited to pipes used in the chemical, oil and gas industries, above all for removal of incrustation. Cleaning agents include acids, lay and special dissolving agents. Pipe must be tightness and its material non attackable by the cleaning product. Neutralization after cleaning and inspection of trace elements in the cleaning chemicals is required.
- *Chemical removal of roots.* Mechanical removal of roots must be regularly repeated as the roots grow rapidly. The alternative solution is to use herbicides in a water solution or in the form of foam. These foams are highly cohesive and adhere to the walls and cavities in the pipes. The vapours are absorbed by the roots, which results in their rotting after about 4 – 20 weeks based on the concentration and contact time. Their effect lasts up to two years. Plant roots can be treated using biological-septic agents, but their use remains sporadic.
- *Biological cleaning methods.* Used to remove grease and fibres from the sewer surface, and for removal of residues and sediments containing phenols, proteins and mineral oils and for removal or suppressing of odours. Components used in hospitals, canteens and food-processing plants. The first component is fatty acids and urea. This liquid is poured into drains and seal-pipes. The pipe system is filled up and hot water added followed by components containing microorganisms. Another method consists in injecting of Sphaerotilus bacteria into sewage water. The bacteria are left in the water for three months. This cleaning technology (using bacteria) can reduce the costs of cleaning.

4.4 THE COST TOOL

One of the most important criteria when choosing a technique is obviously its direct cost. Unfortunately, this becomes also the most difficult task. The difficulties come from the fact that each sewer work is different; the working conditions change, the suppliers and the general construction costs may differ (depending on regions and time). Therefore the aim of assessing the direct cost for rehabilitation technologies is to give a tool for guidance and comparison purposes only.

The Cost Tool presented inside the CARE-S project is based on a few cost drivers that are used by a reduced collection of formulae. Each one of the formulae is suitable for a group of rehabilitation technologies with similar cost behaviour (trench, robotic, man-made, injection, etc.). This tool has an open formulation, so that the end-user can check the default prices and the factors used to assess the cost, and change any of them to adjust it to the local conditions.

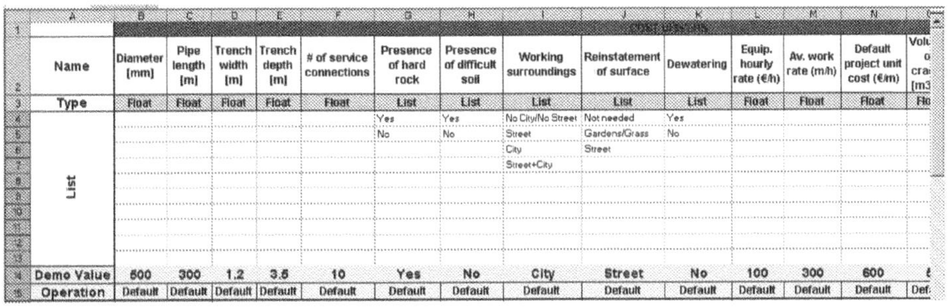

Name	Diameter [mm]	Pipe length [m]	Trench width [m]	Trench depth [m]	# of service connections	Presence of hard rock	Presence of difficult soil	Working surroundings	Reinstatement of surface	Dewatering	Equip. hourly rate (€/h)	Av. work rate (m/h)	Default project unit cost (€/m)	Vol o cra [m3
Type	Float	Float	Float	Float	Float	List	List	List	List	List	Float	Float	Float	Flo
List						Yes	Yes	No City/No Street	Not needed	Yes				
						No	No	Street	Gardens/Grass	No				
								City	Street					
								Street+City						
Demo Value	600	300	1.2	3.6	10	Yes	No	City	Street	No	100	300	600	
Operation	Default	Default	Default	Default	Default	Default	Default	Default	Default	Default	Default	Default	Default	Def

Figure 4.5 Defining the cost drivers

Cost Tool is constructed in an Excel sheet with macros. At a basic level of user, it allows to introduce pipe characteristics in one sheet. These pipe characteristics are pipe identification number, diameter, pipe length, trench width and depth, number of service connections, presence of hard rock or difficult soil, working surroundings (city/rural, street/not), reinstatement of surface, dewatering required, specific equipment hourly rate (if required), average working rate (if different from default), default project unit cost (if known) and volume of cracks. Not all these data are required for all rehabilitation technologies cost calculation. Results are shown in another sheet, which identifies the rehabilitation technology with the pipe ID and two costs. The first cost is for good working conditions, and second is for works in difficult conditions (winter, night, etc.)

For 'advanced users', there is the possibility to open the cost formulation and easily correct the default formulae with their own experience and local market conditions.

REFERENCES

Atkinson, K. 1999. "Sewer rehabilitation and renovation techniques". Subterra developments, UK.

CEN, 1997. "Drain and sewer systems outside buildings". EN 752. EU.

CEN, 2002. "Plastic piping systems for renovation of underground non-pressure drainage and sewerage networks". EN 13566. EU.

Esterková, M. a kol. Bezvýkopová obnova podzemních vedení,GAS s.r.o., Praha

Gue & Partners SDN BHD, 1999. "Specification for Cavity and Slumpzone Probing and Compaction Grout Treatment". Malaysia.

Klepsatel, F. a kol. Bezvýkopová výstavba podzemních vedení, GAS s.r.o., Praha

NRC – CNRC (National Research Council, Canada), 2003. "Selection of technologies for sewer rehabilitation and replacement".

Office of Water, 1999. "Collection systems O&M fact sheet. Trenchless sewer rehabilitation". EPA, USA.

Orman, N. et al., 2001. "Sewerage rehabilitation manual". WRc, UK.

Špalek, P. – Raclavský.J. (2000) Sanace stok a kanalizačních přípojek. Technické doporučení SOVAK, Praha - Hydroprojekt Praha a.s. , 2002

Stein, D. (1999) Instandhaltung von Kanalisationen, Ernst a Sohn Verlag, Berlin, ISBN 3-433-01315-2

Stein, D. 1994. "Rehabilitation of sewer systems". European Water Pollution Control Volume 4, 5 , 1994 , pp49-56

Sterling, R., Simecevic, J. 2001. "Guidelines for pipebursting". ERDC, USACE.

Sterling, R., Simecevic, J. 2001. "Guidelines for Impact Moling". ERDC, USACE.

Sterling, R., Simecevic, J. 2001. "Guidelines for pipe ramming". ERDC, USACE.

Villanueva, A. 2002. "Gestión avanzada de la rehabilitación". CLABSA, Spain.

5

Socio-economic consequences

Jean Philippe Torterotot (*), Delta Sousa e Silva (**), Rémi Barbier (*), Caty Werey (*)with collaboration of Alvaro Pereira (**), Axel König (***), Marielle Montginoul (*), Virginie Waechter (*).

(*) Cemagref ENGEES Strasbourg F (**) LNEC Lisboa P (***) SINTEF Trondheim N

5.1 INTRODUCTION

Rehabilitating sewer networks aims at:

- Complying with regulations and/or levels of serviceability,
- Avoiding or limiting social costs of failures (environment, quality of life, damage, loss of business…) in a sustainable way, while taking into consideration the internal and external costs of rehabilitation works.
- Therefore, there is a need :
    - to define what failures are and to characterise them,
    - o define what the possible social costs are and to understand their links to failures and to works,
    - to analyse impacts of both failures and works on individuals' quality of life as well as on their view about sewerage utility and their level of performance,
    - to assess criteria in order to take the social costs into account for decision making,
    - to develop efficient communication with the public.

These are the topics addressed by work package 5 "Socio-economic impacts of rehabilitation strategies" in project CARE-S.

We could define here social costs of sewer network failures as costs incurred by society (in financial and in non financial terms) and for which utilities or companies have no direct responsibility apart from possible compensation: trade losses of shops, loss of quality of life, repair costs to flooded buildings…

Sewer networks, whatever their physical level of tightness, are open systems; much more open than most network infrastructures. Their inputs (from domestic and non domestic customers, infiltration, surface runoff…), their outputs (exfiltration, overflows, treated water outlet, slums…), but also their location, mostly under public spaces (with consequences in case of failures, repairs and rehabilitation, maintenance), determine this openness.

Hence, different failures and works will have consequences on persons, belongings, activities, which are external to the wastewater utility, and for which this utility may be liable (in legal, financial or moral grounds), even if these consequences may have no or limited counterparts in the utility budget and expenses. To whatever extent this may be, the utility cannot overlook external consequences of its actions or of the network operations.

In practice, the surveys done (the general CARE-S initial survey, with results presented in WP5 report D13, or WP3 survey done by TU Dresden) show that either regulation or management concerns address, to some extent, external consequences of failures or of works. This may be poorly formalised. These consequences may not be assessed or represented as such, but they are underlying to decision criteria or standards.

Inside the CARE-S decision support system, the decision was met to address explicitly at least part of the social costs, which could, in a first approach, be defined as a value given to external consequences of third parties.

Sewer collapses, or blockages, are unanimously considered as failures. When it comes to infiltration or exfiltration, one would probably consider that there exists a threshold (linked to system vulnerability?) above which these unwanted water exchanges are considered failures. When considering flooding or sewerage overflow, the poor condition of pipes (among others collapses and blockages) will increase in probability and/or intensity of events, which may happen anyway; the failure is therefore linked to a hazard increase (flooding or overflow), which may or may not be considered as a failure.

The survey addressed, for 18 utilities, the management frameworks regarding regulation, standards, duties for flooding and overflow events. A majority of utilities have to consider frequency standards, most often based on guidelines or on best practice, with varied flooding return period thresholds. For pollution due to overflows, the discharge standards most often concern the whole sewer system, and compliance of an overflow depends on the situation of this whole system.

5.2 GENERAL PRESENTATION OF WP5 TASKS

Task 5.1 Rehabilitation impact on socio-economic costs

The aim of task 5.1 is to help addressing the socio-economic ("indirect", or "external") costs linked to rehabilitation decision, i.e impacts of failures and impacts of rehabilitation works to third parties, by providing guidance and methods for assessing criteria representing these social costs:

- criteria for comparing a limited set of technologies when considering a given single pipe,
- criteria for comparing various rehabilitation projects, defined each at pipe level, given the related failures or failures hazards.

This task bases on both synthesis of literature (methods, results of studies) and analysis of real world data (claims, compensations, failure events…), and is very strongly linked to work packages 2, 3, 4 and 6. Two software, under spreadsheet format (Socioworks, and Sociofail) are designed and integrated inside CARE-S decision support system, in order to assist the DSS user in assigning values to socio economic criteria.

Task 5.2 Rehabilitation impact on social life quality

This task aims to achieve a systematisation of:

- what the different users perceive as being the impacts of sewer failures and rehabilitation works;
- what are the levels of public acceptability towards failures as well as a set of rehabilitation interventions in their work or residential area;
- what are the public levels of awareness to the importance of maintenance actions, repair and rehabilitation of sewer and storm water networks; this identifies a need to evaluate the extent to which individuals are aware of non-intervention possible risks and consequences related to network failures.

Analysis of these topics is developed through the use of public inquiry methodologies (exploratory interviews, focus groups and questionnaire) in areas previously selected (case studies).

Task 5.3 Public acceptance and communication with public

Based on feedback from field investigations, best practice and guidance are proposed for communication with the public, under circumstances of rehabilitation works or of failures.

5.3 FRAMEWORK AND DEFINITIONS

5.3.1 What are social costs and why consider them?

As mentioned above, sewer networks are open systems. Hence, different failures and works will have consequences on persons, belongings, activities… which are external to the wastewater utility, and for which this utility may be liable (in legal, financial or moral grounds), even if these consequences may have no or limited counterparts in the utility budget and expenses. To whatever extend it may be, the utility cannot overlook external consequences of its actions or of the network operations.

The public nature of wastewater utilities, whatever operated and managed by public or by private structures, is a general fact due probably to the importance of sanitary and environmental stakes, but also due to the natural monopoly of such networks. External consequences are the more a stake for the utility, either managed or controlled by local, regional and/or national government bodies.

In practice, the surveys done (the general CARE-S initial survey, with results presented in WP5 report D13, or WP3 survey done by TU Dresden) show that either regulation or management concerns address to some extend external consequences of failures or of works. This may be not or poorly formalised. These consequences may not be assessed or represented as such, but they are underlying to decision criteria or standards.

Inside the CARE-S decision support system, the decision was met to address explicitly at least part of the social costs, which could in a first approach be defined as the value given to external consequences to third parties. The definitions below have been agreed on for CARE-S project, as parts of the collective glossary.

DEFECT or FAULT

Established condition of the pipe that does not comply with the nominal design and has a lower potential for functionality or use (definition inspired by RERAU – French national project on rehabilitation of sewer networks).

PERFORMANCE DEFICIENCY

Temporary operation or behaviour of the pipe or of the network which does not comply with the general or basic objective; this deficiency is linked to a *defect* and can be linked to an external trigger event such as rain (definition inspired by RERAU – French national project on rehabilitation of sewer networks).

FAILURE

Termination of the ability of a pipe or of a network to perform a required function; a failure is a *defect* or a *performance deficiency* and is defined in reference to a required level of performance (definition inspired by French AFNOR standard on maintenance terminology NF EN 13306).

EXTERNAL EFFECTS (due to defects/performance deficiencies)

Physical changes, due to the *defect* or *performance deficiency*, which impact the "world outside" the undertaking (example: presence of rainwater on a street because the condition of the pipe has decreased the hydraulic capacity, which would otherwise have been sufficient to avoid flooding for this rain event); the "undertaking" is the public or private company or organisation which is in charge of the sewerage service and of the corresponding assets.

EXTERNAL EFFECTS (due to rehabilitation works)

Physical changes, due to rehabilitation works, which impact the "world outside" the undertaking (example: presence of work areas in the street); the "undertaking" is the public or private company or organisation which is in charge of the sewerage service and of the corresponding assets.

EXTERNAL IMPACTS (due to defects/performance deficiencies *or to rehabilitation works)*

Consequences of *external effects* on persons, on activities, on private and public properties and items or on the environment (example: traffic disturbance due to street flooding or to works).

SOCIAL COSTS (due to defects/performance deficiencies *or to rehabilitation works)*

Assessment / valuation of *external impacts*; social costs represent costs incurred by society (including sewerage service customers) as a result of sewerage works or of failures, and for which utilities or companies have no direct responsibility apart from possible compensation; if there is a compensation, the net social costs are the social costs minus the compensation amount (definition inspired by AWWARF).

5.3.2 Impacts of sewer pipes failures: addressing risk and vulnerability

General comments

Literature on risk issues is full of risk definitions. Minimalist risk concepts coexist with broader risk views.

The most common concept of risk is still the one proposed by technical risk analysis, where risk is defined as the product of probability of the occurrence of a specific hazard by the expected value of hazard consequences. Underlying causes tend to be circumscribed to physical phenomena and consequences to tangible damages (number of fatalities, injuries and, in some cases, material damages).

Nevertheless important, technical approaches reveal some limitations. These ones tend to disconsider risk parameters that cannot be easily transformed into a numerical value. Besides, as refers Renn (1992), they assume a mirror relationship between observation and reality and do not consider that causes of harm and its consequences are both mediated by society and human action.

Although belated on the study of risk issues, social sciences had the merit of calling attention to human action interference on the magnitude of, at least, some risks. But, the existence of a plurality of approaches weakens social sciences contribution for a better understanding of risk. Anyway, it should be emphasised that what unifies social sciences diverse approaches is the assumption that the way humans perceive, occupy, transform and manage environment and human activity has an influence on risk severity. The recognition of an environmental hazard as a source of risk is largely dependent on human perception.

Given this, risk configuration depends, a great deal, from the web of relations that link *society* with (built or non-built) *environment,* in time and space. This web of relations dictates the degree of vulnerability of a society towards a certain risk.

In general terms, two major types of risks endanger wastewater systems and their immediate environment: *silent* or *slow manifestation* risks and *sudden* or *abrupt* hazards. Exfiltration and infiltration phenomena are examples of the first mentioned type of risk. Sewer collapses and blockages are examples of the second one.

Above, it was mentioned that vulnerability towards risk varies according to the type of relationship between environment and society. Trying to concretise this assumption to our specific object, we would say that the manifestation of sewer risks is, on the one hand, dependent on event characteristics and, on the other, on built environment and societal characteristics. They can be synthesised as follows:

- characteristics of the event: magnitude, intensity, timing and duration
- "hydrological" environment and network characteristics: climate and its contingencies, urban catchment, wastewater and storm water type of infra-structures and management
- land-use and population characteristics: demographic and socio-economic population characteristics, risk perceptions and protective behaviours.

Conceptual scheme

The scheme below displays the concepts introduced with the above definitions. Vulnerability has not been introduced in the definitions: we will define it as the quantity of "stakes" which can be exposed to external effects (population, goods, activities, natural patrimony and resources...) and the fragility of it towards the effects.

Let us consider three examples:

(1) The pipe is not tight, and there is exfiltration.

We will consider this exfiltration as a continuous phenomena, even if during rains the increased water level inside the pipe may increase exfiltration. The defect is non-tightness, and the misfunction, which exists without an external trigger, is exfiltration. The external effect will be wastewater released in the soil and possibly in the groundwater. The pollutant transfer effect will depend on exfiltration discharge, soil type (permeability) and distance to groundwater. The initial quality of the groundwater, the global volume and dynamics (flow speed, renewal period) and the present or future water uses (which may be sensitive to this pollution) will be the parameters accounting for vulnerability. The consequences on present and future water uses will be the external impacts, and the valuations of these impacts are the social costs.

(2) The pipe has a reduced hydraulic capacity.
This means that during any rain, the water level inside (or outside) the pipe will be higher upstream from that pipe, and that there will be overflows, water overtopping in basements or on the surface, more that if the condition of the pipe was good; the differential in hydraulics as soon as it shows outside the network will be the effect of the misfunction. People, goods, activities and traffic in the area or in the buildings which will be more flooded than "normally", together with their sensitivity to flooding, will constitute the vulnerability. All types of damage or disruption will constitute the external impacts.

(3) The pipe has a structural collapse.
This defect may have several consequences, for instance in terms of flooding or odours. Let us consider the case where this collapse leads to surface soil depression; the defect (collapse) has a direct external effect - soil depression. The vulnerability of the pipe environment will depend on soil use, traffic (if it is a road) and external impacts may, according to the situation, cause damage to road structure or at least (increased) traffic jams.

One crucial point is that we should logically only take into account those defects and malfunctions which are to be considered as failures. Minor blockages in separated wastewater pipes may not be failures, though they are defects. A flooding event due to rain which exceeds the network design and the regulation or standards, and which is not increased by the hydraulic operation and condition of the network, is not a failure of the network.

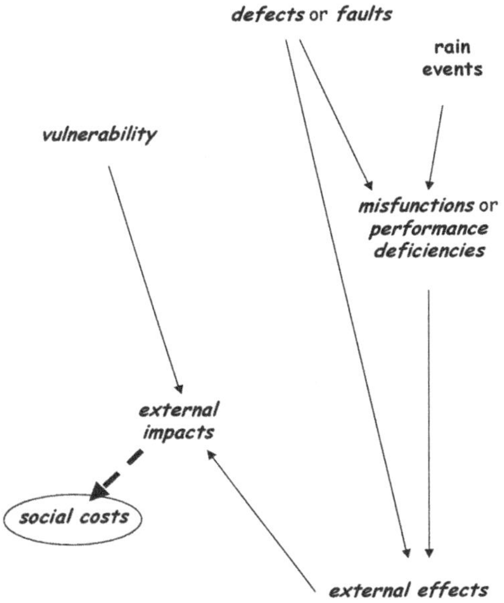

Figure 5.1 Concepts of failures and related social impacts

Definition of failures

Let us recall the definition proposed for "failure": *termination of the ability of a pipe or of a network to perform a required function; a failure is a defect or a performance deficiency and is defined in reference to a required level of performance.*

A defect or a misfunction would be a failure if they do not comply with:

- the regulation and compulsory standards
- voluntary standards
- the initial design and foreseen operation conditions
- the will of the utility manager / utility owner

The non compliance can be defined, measured, characterised through:

- the existence or occurrence of the defect or misfunction (for instance a structural pipe collapse)
- the frequency of the defect / misfunction (for instance a combined overflow exceeding the permitted frequency); The Middlesex University Flood Hazard Research Centre (Green et al 1993, Tunstall et al 1993) performed a survey for OFWAT on the perception of water and sewerage customers: the majority of people ask for being exposed to less than 2 network flooding every 10 years. Flooding is generally considered worse than having a car stolen, and as serious as a fire in the kitchen with the fire brigade being obliged to intervene.
- the intensity of the defect / misfunction (for instance non tightness leading to infiltration exceeding a given percentage of dry weather flow)
- the occurrence of the misfunction given the frequency of the trigger event (for instance flooding occurring for a rain of small return period); it would be probably accepted that a network overtops with a 100 years storm if the network is designed for a 10 years storm, and provided the condition of the network does not make this flooding significantly worse
- the consequences of the defect / misfunction (for instance odours due to small blockages).

Annex 1 of report D13 (Werey et al, 2005) shows the results of the internal CARE-S survey concerning regulations on performance, standards … regarding failures. There is a major diversity of situations. Defining the boundaries of failures is a local issue, at undertaking and local government levels. Even if there are national standards, the wastewater service may decide to “do better” and to be more self-demanding towards failures.

External impacts are consequences of failures effects on third parties. But this does not mean that these third parties bare all the impacts. If the wastewater service pays for compensations, this reduces the social cost suffered by these third parties (hence the “net social costs” representing the “remaining part”), and increases the internal cost for the undertaking. If compensations are covered by the undertaking’s insurance, it will also reduce the net social costs and increase indirectly the internal costs, as the undertaking has to pay for the insurance contract (on a “permanent basis” and not for each failure).

For exceeding pollution discharge, the undertaking may also pay for pollution taxes (according to the polluter pays principle), this money being used to enhance the aquatic environment. In that case, we should consider that the net social costs of the pollution are reduced by the amount of pollution taxes, and that these taxes are part of the undertaking internal costs. In case of a “fine”, which would not be used for the sake of water resources, the net social costs would not be reduced (as the pollution is not made for), but the internal costs increased.

Annex 1 of report D13 (Werey et al, 2005) gives an idea of the diversity of situations among countries and CARE-S end-users, as far as compensations are concerned. The difference between social costs and net social costs will depend on each local regulation or policy. We should therefore address “whole social costs”. But on the basic principle, net social costs excluding direct or indirect compensations should be taken into account. If these compensations where fair and comprehensive, they could be a good assessment for social costs.

5.4 TASK 5.1: REHABILITATION IMPACTS ON SOCIO-ECONOMIC COSTS

In an ideal world, defining and assessing social cost criteria would be based on a good knowledge and description of failures or works, of the external effects (flooding conditions, soil sinking in due to sewer collapses…), of the pipe environment (street characteristics, traffic, buildings…), and would also consider appropriate economic assessment methods which require also specific data. In practice, various points may be insufficiently known, and uncertainties are sizeable.

As the criteria are to be used in multi-criteria selection procedures, the objective is to define and assess criteria which are consistent with the social costs, whether calculated in monetary terms or not, and which allow to deliver a sound and unbiased comparison of either two different rehabilitation projects, or two different rehabilitation technologies, in terms of avoided social impacts of various kinds. Keywords here are feasibility, consistency and possibility by the utility manager to consider criteria definition, which make sense for him.

5.4.1 Structure of the task and relations to other CARE-S components

The aim of WP5.1 is to assist the user of the CARE-S decision support system in defining criteria which allow taking into account, directly or through coherent indicators, socio-economic costs of external impacts of either network failures (WP5.1b, Sociofail software) or rehabilitation works (WP5.1a, Socioworks software). These criteria can be used for decision making within WP6 (SRT and SRP, developed within WP6), and uses results from tolls developed within WPs 2, 3 and 4.

WP5.1 generic inputs/outputs and the way WP6 and WP5 procedures combine are explained in detail in D13 report (Werey et al 2005).

Criteria for the selection of priorities among rehabilitation projects

This procedure considers a pre-selected set of rehabilitation projects, where each project is a pipe defined with its characteristics (intrinsic ones, local environment…) and data on possible failures.

As mentioned above, WP5.1b criteria are meant to help ranking pipes candidate for rehabilitation, considering their potential failures. These criteria, representing socio-economic costs of external impacts (definitions given in the next section) must hence take into account the potential effects a failure can generate, and the vulnerability of the concerned pipe environment to such effects. If an area shows no vulnerability to a given effect, in other words if there are no impacts, the criteria must show "no cost".

The occurrences of some failures are of probabilistic nature, such as wet weather flooding or blockages. This probabilistic nature must be taken into account, in a risk assessment approach for the criteria. Some failures may generate "randomly" external effects, such as soil; surface depression due to collapse, which cannot be assessed in probabilistic terms. For such cases, decision was met to consider "threat factors" criteria defined as the probability of the initial failure (e.g. pipe collapse) times a vulnerability indicator.

WP2 and WP3 will allow the consideration of the evolution of pipe condition and of potential failures. What about assessing future socio-economic costs through WP5.1b criteria? The decision met was to consider only the present situation, for the following reasons:

- though the socio-economic environment of a pipe may change, in average, slower that pipe condition, when changes occur they can be quite drastic (replacing an old industrial area by collective housing, having a 4 lanes connexion road created instead of a small street…); in any case, forecasting pipe environment changes needs case by case investigations and is less deterministic than the random evolution of pipes condition;
- rehabilitation planning considers time horizons for which land uses, population densities, road characteristics … are stable out of focused drastic changes like mentioned above.

What are the failures, effects and impacts considered?

Possible defects, misfunctions and external effects are numerous and varied. So are failures. We can only address defects and misfunctions for which there is information available about their possible occurrence, intensity, and/or probability, and which are to be considered as failures. Moreover, the question is then to link the external effects with their occurrence / characteristics / intensity / probability to the failure events, and the impacts to vulnerability and to effects. Failures can only be taken into account as far as they are addressed and described, to whatever extent, in "technical" terms in WP2 and WP3 outputs.

For the effects, which are of probabilistic nature, we have to consider the notion of risk: probability times consequences, the latter being the crossing of effects intensity (water height…) and vulnerability. Effects can also be, to some extend, "random" towards defects or misfunctions:

- blockages may or may not generate dry weather wastewater overtopping in basements or on the soil surface
- blockages and collapses (and more generally structural failures) may or may not induce problems of odours, insects, rodents
- structural pipe collapses may or may not induce soil depression.

In that case, it is not possible to address the chain [defect or misfunction → external effects → impacts], in a deterministic or even full probabilistic way. As will be seen later, the choice was met to consider "threat factors", combining the probability of the "trigger failure" (blockage, collapse, structural failure) with a vulnerability indicator to the possible external effects (overtopping odours and soil depression) : for instance, this would mean combining probability of pipe collapse with road traffic. This must in no way be considered as a "risk", as we miss the probability of the damaging effects. But such a "threat factor" combines the "best available" information accounting for the potential risk situation.

The table below displays the failures, effects and impacts taken into account in WP5, and described in more detail in D13 report (description, existing information and data, proposition for a decision criteria inside the CARE-S framework). The numbers relate to the types of impacts addressed in WP5 and for which criteria are considered.

Table 5.1. Failures, their external effects and socio economic impacts considered in decision making about prioritising pipes for rehabilitation

	DEFECTS AND MISFUNCTIONS					
		Wet weather floodings (with probability)	Blockages (with probability) in dry weather situation	Structural pipe collapses (with probability)	Combined sewer overflows	Exfiltration
	Are they considered as failures?	Choice of end-user according to probabilities of reaching given levels	All of them	All of them	Choice of end-user according to a set of parameters	Any
	Through outputs from	WP3 (hydraulics) considering structural condition (WP2)	WP2	WP2	WP3 considering structural condition (WP2)	WP2
E	Presence of water in basements	1, 2	8			
F	Presence of water on soil surface	1, 2, 3	9			
F	Presence of water on buildings ground floor	1, 2				
E	Pollution discharge to surface receiving waters				4	
C	Pollution of groundwater (defined through "groundwater vulnerability" WP3)					5
T	No discharge possible to sewer		6			
S	Soil depression			7		
	Odours, insects, rodents		10	10		

1: flooding damage to buildings and their contents, and business losses
2: flooding intangible damage
3: traffic disruption and trouble due to flooding
4: degradation of surface water quality and consequences on present or future water uses
5: degradation of groundwater quality and consequences on present or future water uses
6: wastewater service interruption
7: traffic disruption and trouble, annoyances to life quality, due to soil depression
8: annoyances and damage from wastewater overtopping in basements
9: annoyances and damage from wastewater overtopping on the street
10: annoyances due to odours, insects, rodents.

Among the failures not considered for socio-economic impacts, we should mention wastewater treatment plant performance reductions due to clean water infiltration. Two reasons account for not considering it:

- this phenomenon, more than any other considered here, is a cumulative one over the whole network upstream from a wastewater treatment plant; identifying and quantifying the contribution of a single pipe is quite cumbersome;
- the effect on the receiving water is due to the operation of the whole sanitation and treatment system, which can be assessed, but makes it the more difficult to identify a quantitative or ranking criteria.

Criteria for the selection of a rehab technique for one pipe

This procedure considers a pre-selected set of "feasible" and "acceptable" rehabilitation technologies (described with various characteristics) for a given pipe, defined with its intrinsic and environment characteristics.

As mentioned above, WP5.1a criteria are meant to help ranking rehabilitation techniques, which could be applied to a given pipe. These criteria, representing socio-economic costs of external impacts (definitions given in the next section) must hence take into account the potential effects a technique can generate, and the vulnerability of the concerned pipe environment to such effects. If an area shows no vulnerability to a given effect, for instance noise, in other words if there are no impacts (for instance area with no population and no established labour force), the criteria must show "no cost", and hence no difference to be made between two technologies whatever the noise created. If two technologies generate different levels of effects and consequently of impacts, for instance road surface neutralised, this must make a difference on the "traffic disturbance" criteria.

What are the effects and impacts considered?

These situations are quite simpler than failure ones, as we consider rehabilitation works with different techniques (as described and characterised by WP4), which will happen in a given location (a pipe with its socio-economic and physical / biological site environment). The possible effects are those addressed in the WP4 database, directly or indirectly:

- noise
- dust
- groundwater pollution hazards
- service interruption
- external surface neutralisation (digging, working areas) leading to traffic disturbance and annoyances, and losses of business independently from traffic consequences (smaller attractiveness of shops...).

D13 report addresses each type of impact related to this list (description, existing information and data, proposition for a decision criteria inside the CARE-S framework). The same conceptual description in terms of external effects, vulnerability and external impacts can be done as for failures, but in a more usual and simple way. The works are a certain event, and we can consider that all (or the major part) of the effects are quite deterministic.

The consequences to people (inhabitants, service customers, non residents like labour force and by-passers), to activities and traffic, to groundwater resources and uses are addressed in terms of external impacts.

The duration of works will have an obvious influence on the external impacts. Inside the WP5 procedure, based on information available on pipe (length) and techniques, the DSS user will be able to assess the duration for each technology, by considering among others the linear performance of the rehabilitation technology.

5.4.2 The Sociofail and Socioworks softwares

Sociofail

General description

"Socio-fail" is an Excel® sheet based software, part of the CARE-S decision support system, which is meant to prepare criteria related to socio-economic costs of sewer pipes failures. These criteria are used by the CARE-S multi-criteria procedure that helps to prioritise the pipes for rehabilitation (CARES-SRP: selection of rehabilitation projects).

The criteria try to give an unbiased representation of the socio-economic costs, given the information available for the pipe considered and for the failures. They are specific to the CARE-S procedure, as they depend on the way pipes and failures are described in this decision support system.

The software user has the possibility to validate or change every data related to the environment of the pipe, as well as any criteria value. The user can choose the criteria values according to his/her needs, and take into account his local knowledge of socio-economic costs due to sewer failures.

Scope of application

The tool is launched by the CARE-S rehabilitation manager, and considers: these ones have been previously pre-selected by another procedure, and are described with potential / observed / estimated failures in their present condition through parameters taken from the corresponding CARE-S databases and tools.

Pipe and failures related data that represent meaningful information for the possible consequences of failures are displayed. Default criteria calculations are proposed, based on a minimal view concerning available information. If such data are missing, there are no criteria calculated by the tool. The tool user has the possibility to validate and change the criteria values, according to his knowledge and to local information available.

Output data

There are two types of output data:

- the pipes environments data which may have been modified or completed; they will be stored with their possible new values in the general CARE-S database corresponding to the current data set;
- the criteria values for each pipe and for each type of cost or impact.

Socioworks

General description

"Socio-works" is an Excel® sheet based software, part of the CARE-S decision support system, which is meant to prepare criteria related to socio-economic costs of sewer pipes rehabilitation works. These criteria are used by the CARE-S multi-criteria procedure, which helps to choose the best rehabilitation technology for a pipe, which has been previously selected as one of the priority pipes (CARES-SRT: selection of rehabilitation techniques).

The criteria try to give an unbiased representation of the socio-economic costs, given the information available for the pipe considered and for the rehabilitation technologies. They are specific to the CARE-S procedure, as they depend on the way pipes and technologies are described in this decision support system.

The software user has the possibility to validate or change every data related to the environment of the pipe, as well as any criteria value. The user can choose the criteria values according to his/her needs, and take into account his local knowledge of socio-economic costs due to rehabilitation works.

Scope of application

The tool is launched inside the CARE-S decision support system, and considers one pipe by run, as well as a varied number of possible rehabilitation technologies: these ones have been previously pre-selected by another procedure, and are described through parameters taken from the corresponding CARE-S database.

A crucial characteristic of rehabilitation works, regarding their external socio-economic consequences, is the duration including installation and dismantling of the work areas and related appliances, as well as cleaning of the area. A default calculation of duration is proposed according to the length of the pipe and to the linear speed of the technology. Such a default calculation is not only rough, but also incomplete. The tool user can and should validate and correct this value, according to the technology, the specific characteristics of the site, and the way works may be organised. This duration is used for most of the criteria calculations.

Pipe and technologies related data, which represent meaningful information for the possible consequences of works are displayed. Default criteria calculations are proposed, based on a minimal view concerning available information. If such data are missing, there are no criteria calculated by the tool. The tool user has the possibility to validate and change the criteria values, according to his knowledge and to local information available.

Output data

There are three types of output data:

- the pipe environment data which may have been modified or completed; they will be stored with their possible new values in the general CARE-S database corresponding to the current data set;
- the duration of works for each pre-selected technology;
- the criteria values for each technology and for each type of cost or impact.

Strengths and limitations of the tools

This tool allows the user either to consider default criteria proposed, or to define his/her own criteria according to his knowledge, to his/her interpretation of literature results given in the accompanying report, to local field information. Even if the user wishes to enter new values, the tool allows the user to manage input and output, and to display relevant information related to the pipe, its environment, and the technologies. This may in particular be useful for dealing with areas / streets which have specific and important vulnerabilities to works. Criteria must not only be "logically sound", they must be meaningful to the tool user and decision maker. In that sense, the possible "do it yourself" use of the tool is an advantage.

The drawback of the principles recalled above, is that defining the criteria may be time consuming, as the default calculations produce very basic criteria with limited socio-economic criteria, which are more indicators than assessments in the usual meaning.

5.5 TASK 5.2: REHABILITATION IMPACT ON SOCIAL LIFE QUALITY

5.5.1 Introduction

Knowledge about impacts of both sewer failures and rehabilitation works implies to consider the way modern societies cope with sewage, with special attention to the nature and mode of relationship between sewer systems and users.

Therefore, this task of the project stands on a theoretical framework, which views sewer systems as one in an amount of *expert systems* that colonise modern industrial societies. These are physical artefacts or systems of technical realisation, whose conception and management involves expert knowledge and expertise (Giddens, 1990). The vitality of such systems to social quality of life contrasts with its compulsory and unquestioned use. In fact, individuals are fundamentally lay users of these types of systems. Their relation with sewerage-*expert system*, including with who manages it, is not based on any kind of expert knowledge, but on a *trust* or belief on those systems' reliability. Such non-reflexive *trust* stands on the daily confirmation, through experience, of systems' normal functioning as well as on a kind of faith on the good quality of expertise involved on it.

Failures and rehabilitation works are, in this context, envisaged as *critical events,* that is abnormal situations which may introduce breaches in the expected course of daily life and in the expected operation of the sewerage-*expert system*. It was hypothesised that such breaches may express themselves at two levels: individuals *quality of life* and the relation *users ↔ sewerage utility*, based on *trust*.

Concerning individuals *quality of life*, we postulate that negative impacts may be felt in the areas of *housing*, *health & well-being*, *near environment* and *works & finances*. Attached to this, it appeared as pertinent to explore the degree of public acceptability towards such events.

Breaches in the relation *users ↔ sewerage utility* renders concrete mainly through a decrease of customers' *trust* on services. The way utilities manage such type of *critical events* may play an important role of slowing down *trust* erosion.

This framework is summarized in Figure 5.2, and has been used to structure the analysis.

5.5.2 Methodology

Present study stands on a case-study type of methodology, which comprehends two target-areas: Amadora-Oeiras municipalities (Lisbon-Portugal) and Nantes Metropole (urban community in France). On both areas two interrelated studies were pursued, a first one of exploratory nature and a second one, conceived to deepen empirical knowledge and extend data collection to a large number of individuals.

Below, main techniques and procedures underlying both studies are summarised.

Exploratory study

Besides its role of first approach to the field, exploratory study aimed to be a support tool for the survey preparation.

This study stood on two procedures: semi-directed interviews to *privileged informers* and *focus groups* discussions with victims of sewer failure events and individuals with experience of rehabilitation works.

Interviews were done with people representing sewerage utility on both areas, other local officials, representatives of population (associations...). *Focus groups* sessions were organised, -making an effort of gathering people representative of the diversity of inhabitants and *professionals*, working in the area.-.

Along with the role of supporting tool to the further step of data collection, fieldwork pursued under the exploratory study allowed a contact with vocabulary and other relevant issues that are meaningful for the people to survey.

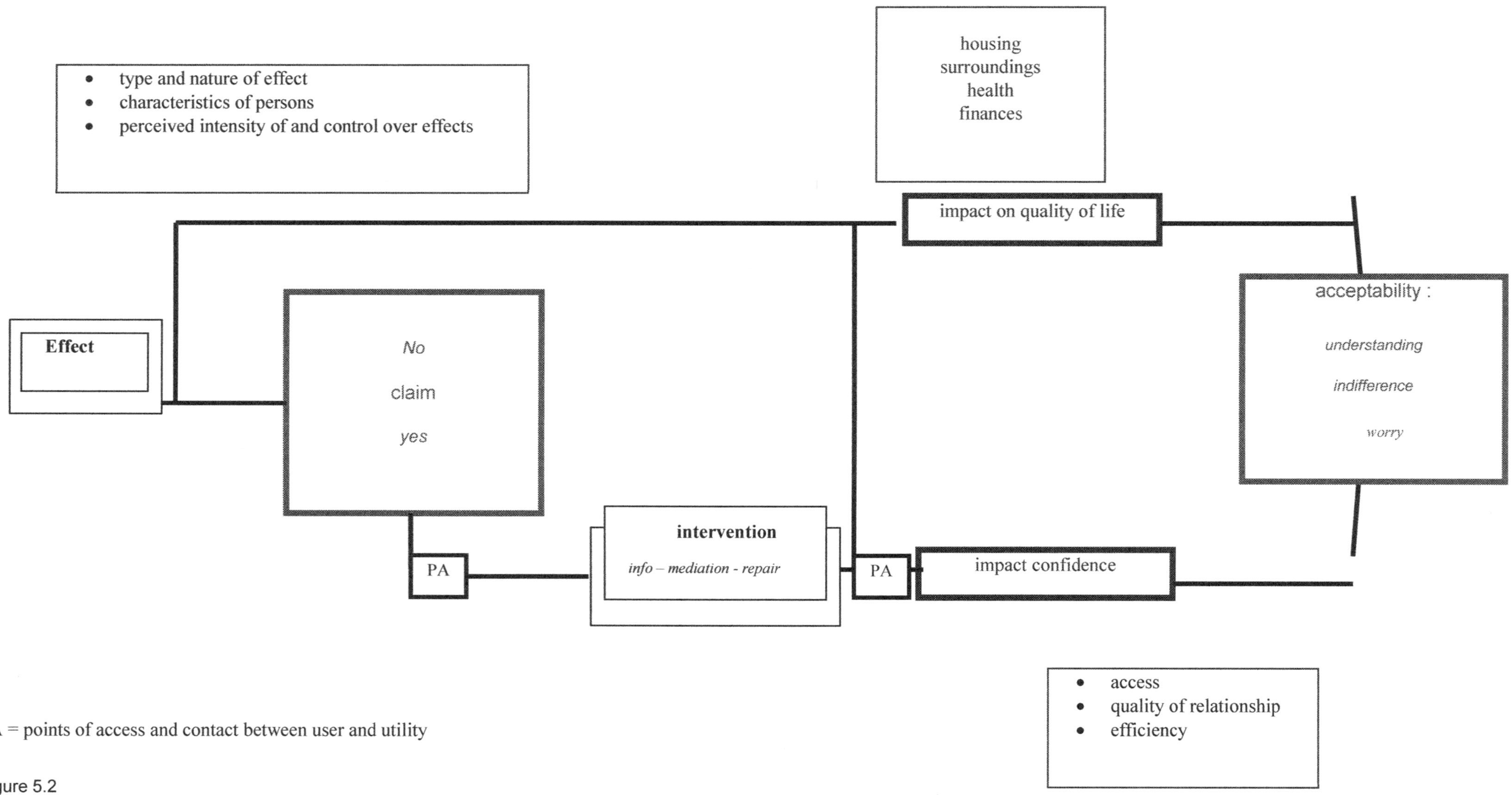

PA = points of access and contact between user and utility

Figure 5.2

The survey

The survey type of approach implied the development of a set of procedures, with the questionnaire construction and sampling on both target areas. This was done in the light of a theoretical framework

The process that gave birth to the questionnaire followed an exercise of decomposition of concepts and propositions into dimensions and indicators, having research hypothesis as background. The approach of data collection that emerged from such a process was composed by a set of questions around the following dimensions or parameters:

- nature and description of critical events and of their effects;
- perceived impacts on quality of life (housing or work place, health & well-being, near environment-, work & finances);
- acceptability towards *critical events*;
- experience of contact with the sewerage service, expectations towards sewerage services and *trust*;
- general information (socio-economic characteristics of persons, knowledge and understanding of sewerage issues…).

Considering the aims of this study research, two types of individuals appeared as fundamental to integrate on the sample. We refer on the one hand to victims of sewer failures and, on the other hand, to individuals living or working on areas recently object of rehabilitation works. Apart from these, a third target-group was introduced, that is individuals without experience of any of those events. This one intended to function as a control group along data analysis.

Given the means available, and the diverse numbers of people who might be surveyed in different groups, it was not realistic to design a statistically representative sample. Therefore, the option was of following a non-probabilistic and purposive type of sample procedure. The table below shows the target numbers of interviews, and the total numbers of interviews, which were done and analysed.

Table 5.2 Sampling considered initially for the surveys, and total numbers of interviews done

	Failures			Sewer rehabilitation works			Non-victims	
	Struct.	Occasi.	TOTAL	Epicent.	Surroun.	TOTAL		**TOTAL**
Residents	20-25	20-25	40-50	20-25	20-25	40-50	20-25	100-125
Professionals	15-20	15-20	30-40	15-20	15-20	30-40	15-20	75-125
TOTAL	35-45	35-45	**70-90**	35-45	35-45	**70-90**	30-40	**170-220**

Oeiras – Amadora sample

	Failures	Rehab works	Non-victims	TOTAL
Residents	52	50	22	124
Professionals	33	30	14	77
TOTAL	85	80	36	201

Nantes sample

	Failures	Rehab works	Non-victims	TOTAL
Residents	52	49	30	131
Professionals	39	30	19	88
TOTAL	91	79	49	219

As can be seen through the tables above, the sample was composed by residents and an amount of individuals whose main bound with the target-area is of professional order (shopkeepers or other). Besides, it was made an option of assuring, as far as possible, the presence of victims of structural and occasional failures as well as the presence of individuals living in the epicentre of works and on the surroundings.

With the support sewerage services and their staff, people or areas to survey were previously identified either on the basis of recent works, recent failure events, or individual claims corresponding to very local or individual cases.

The questionnaires were applied by a group of trained interviewers, at Nantes and in Oeiras-Amadora target areas. The great majority of them followed the procedure of face-to-face interviewing.

After validation and intake of the questionnaires in databases, the results were analysed, privileging bi-variate and multivariate types of analysis approaches.

5.5.3 Some key synthetic results

The different approaches for analysing and surveying the perception of population and acceptability should not be summarised into a too short list of statements. Nevertheless hereafter some key points are listed, which gives a simplified view of the outcomes.

The two surveyed sites showed to be quite different in terms of the types of the most frequent failures occurring. The results of the population and professionals surveys also showed some differences. Common points appeared such as:

- rehabilitation events are more easily accepted by the people than failures;
- impacts which cannot be controlled are felt as less acceptable; intensity of an impact has only a secondary influence;
- during a failure event, people expect, from the utility, more efficiency of intervention than speed of intervention.
- More detailed conclusions are listed below for each survey site.

Lisbon case

- Sewer failures induced on impacts for almost every individual. A half of them felt such consequences as significant to their own quality of life. Nearby surroundings and housing/workplace were the areas of quality of life where impacts were felt as more disturbing. Such tendency is, in fact, congruent with the nature of failure. Most interviewees were victims of sewer flooding.
- Experience of sewer rehabilitation works seems to have induced on modest impacts to Lisbon sample subjects. Problems of use and usufruct of nearby surroundings is what seems to have disturbed more rehab' works subjects. Attention must be paid to the *time slowing down effect* . Target-areas for rehab works inquiry were areas where works had occurred long times ago. Given this, it is probable that felt impacts had vanished from some individuals' memory.
- Sewer failure type of *critical event* is mainly perceived as an unacceptable event that should definitely be avoided. Concerning rehab works type of *critical event*, Lisbon subjects revealed higher levels of acceptability or tolerance.
- The number of individuals who turns to wastewater service, asking for support or to claim, is limited. Failure victims declared not to know whom to contact, and that they didn't feel like rehab works' subjects would need doing so.
- Efficiency on problem solving is the aspect that emerges as the most important to individuals, especially failure victims, even more than speed of response.
- *General trust* on wastewater service varies according to the target-group and *critical event*. The tendency is for failure victims to distrust services; whereas rehab subjects reveal some trust. Experience of failure victims with the services (*access points)* tends to perpetuate or provoke a situation of distrust towards sewerage service performance.

- Lisbon case study reveals that experience of *critical events*, especially sewer failures, seems to have roused more awareness towards the importance of sewer rehabilitation.

Nantes case

- Users expectations are a hidden and confined operation of sewer systems (no odour, no noise, nothing in sight), which is smooth and continuous. The expectation about a good and smooth operation of the sewers relies on a general feeling of confidence of users towards the "sewerage expert system".
- The more intensively perceived impacts concern in-house annoyances and domestic quality of life.
- Among these "domestic impacts", the more disturbing or unaccepted ones are those which cannot be easily controlled (noise, odours), ie which may leave both users and the sanitation department powerless, at least in the short term.
- Notwithstanding the impossibility of taking definite conclusions, data analysis indicates that Nantes subjects reveal, at the level of the discourse, some awareness towards the environmental importance of sewerage.
- Acceptability of critical events has only a secondary dependence on perceived intensity of impacts. Indeed, in spite of being mainly perceived as events of low significance, sewer failures are viewed as unacceptable or intolerable events by a non-negligible amount of Nantes interviewees. Concerning rehabilitation works, there is an amount of individuals who perceived this event impacts as having been disturbing for their lives. Nevertheless, they seem to be envisaged as bearable situations. Such trend indicates that the type of *critical event* influences more acceptability, than level of impacts' intensity or degree of disturbance. As far as it is possible to infer, there is apparently more predisposition to bear negative consequences if individuals anticipate some kind of benefit.
- Critical events question the whole "municipality citizen" relationship, which designate both the municipality as the service provider organisation (the service relationship around the sewerage system), and the municipality as the guarantor of social unity for local society (community relations).. As for community relations, they seem to us to be underpinned by two essential qualities of the inhabitants: voter and taxpayer. Therefore, behaviour that is apparently at odds with the efficient and thrifty use of "public money" will be criticised.
- In cases of failures, the service relationship is questioned according to the level of apparent efficiency of the intervention, i.e the degree to which the failures effects have been controlled by the intervention.
- In summary, the expectations of users towards the utility concern mainly three topics: technical efficiency, speed of intervention, ease of contact to the utility.

5.6 Task 5.3: Public acceptance and communication with public

This task aims at proposing recommendations, for communication and interface with the public in connection to sewers failures and rehabilitation. It is based on:

- the analysis of social accompaniment procedures in Oeiras and Amadora, Portugal (task 5.1), and the users policy developed in the Urban Community Nantes, France (task 5.2),
- the preliminary analysis and population surveys in the Lisbon and Nantes urban areas (task 5.2), and the analysis of the results related to the relationship between users and utility,
- a more general reflection on the way present modern societies manage sewerage utilities.

With sustainability challenges, such as ageing of assets, sewerage utilities need to "get the user (back) inside the system", or in other terms to "re-convoke" him. For the issues dealt with by CARE-S, the user is concerned as a possible victim of bad conditions of assets, as well as the payer of protection, prevention and repair costs. He is therefore in a central position amongst the trade-offs to be made. The question of understanding the stakes and of accepting them or not is of growing importance.

Moreover, there is a broad movement of growing expectancies of users towards the quality of the service delivery: users see themselves both as users and as customers. Local services have to design and implement "user's policies" based on some key-words as quality, proximity, innovation.

We propose that, in case of failures, interaction with the public should follow a strategy that would privilege the following dimensions:

- ***accessibility*** to sewerage utility, which means that the way of approaching the sewerage utility in case of need should be an easy one, diverse and previously known by customers;
- ***commitment*** on solving customer's problem in a reasonable time;
- delivery of proper ***explanation*** about what happened, why it happened and how the utility solved or intends to solve the problem (OFWAT,2001).

In what concerns to rehab works, the first effort should be the one of anticipating potential impacts and limiting them through proper planning. We propose that, in case of works, interaction with the public should follow a strategy that would privilege the following tools:

- the ***meeting*** with the population before works start, which is an opportunity to establish a direct contact and dialogue with users and to listen to their possible specific concern;
- a procedure to deliver ***pamphlets*** which contain information awaited by the public (Planning : date of start and estimated duration; Nature and benefits of the works ; it's both possible and useful to emphasis simultaneously on benefits for the users and for the environment ; Possible consequences: traffic and parking problems; access to house / shop; odours problems; WC interruption… ; Whom to contact in case of question / problem (phone number) ;
- a process of ***mediation*** along works occurrence aiming at solving unexpected problems and at avoiding critical practices ; according to the experience of Nantes Metropole, a mediator may intervene for example to check out that companies do respect their schedule or to help people managing their access / exit problems… More generally, a contact officer can help creating a "good atmosphere" between all stakeholders.

All suggestions have to be adapted to the importance and duration of the work and to the technology used.

References

Barbier R. and Sousa e Silva D. (2005). Public acceptance of rehabilitation action and communication with the public, CARE-S report D15.

Giddens (1994). Les conséquences de la modernité. Paris: L'Harmattan.

Green, C. H., S. M. Tunstall, M. Herring and J. Sawyer (1993). Customer preference and willingness to pay for selected water and sewerage services - technical report to the Office of Water Services. Enfield (United Kingdom), Middlesex University Flood Hazard Research Centre: 136.

OFWAT (2001). Understanding Costumers Views: an OFWAT/ONCC Report. Birmingham: United Kingdom.

Renn, O. (1992). 'Concepts of risk: a classification'. Em S. Krimsky e D. Golding (orgs.). *Social theories of risk.* London, Praeger, pp. 53-79.

Sousa e Silva D., Barbier R. and Montginoul M. (2005). Rehabilitation impact on social life quality, CARE-S report D14.

Tunstall, S. M., Green C.H., Sawyer J. and Herring M. (1993). Customer preference and willingness to pay for selected water and sewerage services - summary report to the Office of Water Services. Enfield (United Kingdom), Middlesex University Flood Hazard Research Centre: 27 p. + ann.

Werey C., Torterotot JP., Sousa e Silva D., Kønig A., Peireira A. and Montginoul M. (2005). Rehabilitation impacts on socio-economic costs, CARE-S report D13.

6

Multi-criteria decision support

Rolf Baur and Raimund Herz, Technische Universität Dresden

6.1 BACKGROUND

The ageing process of pipelines in sewer networks leads inevitably to a loss in performance and a risk to the environment, be it through blockages, sedimentation, root intrusion and subsequent hydraulic problems, through bottlenecks leading to surface and basement flooding, and frequent discharge of untreated sewage into the receiving waters. Thus, as more and more sewer pipelines are deteriorating, wastewater companies are facing rising rehabilitation needs in their networks. CCTV inspection is the common approach to monitor the condition of sewer pipelines and to take appropriate measures as soon as they stay in a critical state. In order to guarantee the serviceability of sewer and storm water networks operators should rehabilitate their systems pro-actively, i.e. before parts of the system loose their functionality. The functionality of the system includes hydraulic aspects, environmental aspects, and structural and operational aspects. Furthermore, it is the objective of the system's operator to manage the rehabilitation of the network in a cost-efficient way. The measures taken are subsequently monitored again with respect to the effects they have on the performance indicators (Baur/Herz 2002, Baur/Kropp/Herz 2003, Schulz/Baur/Krebs 2004). The CARE-S project aims to support this objective by the development of procedures for the identification of rehabilitation projects with highest priority, for choosing the most appropriate rehabilitation technology of these projects, and for the development of a long term rehabilitation strategy in order to keep the rehabilitation budget steady and to meet the utility's objectives of serviceability. Compared to other underground infrastructure, it is an advantage that sewer systems can be optically inspected to a large degree under operation. A main issue of a decision support system will be the beneficial use of these inspection results for an efficient rehabilitation planning. Within CARE-S, the definition of decision support methodologies for strategic planning of rehabilitation investments, selecting priority pipes for rehabilitation, and choosing the appropriate rehabilitation technology or measure was dedicated to working package 6. Accordingly, this chapter is structured in four main sections: Section 6.2 presents a classification of methodologies of multi-criteria decision support. Section 6.3 deals with the application of the sewer network deterioration model developed in WP2 for the calculation and evaluation of long-term rehabilitation strategies; in section 6.4 the implementation of a rehabilitation strategy into a rehabilitation plan by setting priorities is discussed. For a selected pipe, the procedure for choosing the best rehabilitation technology is applied in section 6.5. The work has been subdivided into the following three areas of research:

(1) Exploring rehabilitation programmes and strategies (section 6.3)

Long-term rehabilitation programmes and strategies have to be developed on the basis of a condition forecast of the local network. Key indicators associated with the condition forecast of the actual stock are compared to

 ISBN: 9781843391159. Published by IWA Publishing, London UK

their development for different rehabilitation programmes and strategies. The rehabilitation strategy analysis will show the expected financial consequences of specific network rehabilitation programmes and their impacts on technical and economic performance indicators.

(2) Selecting cost-efficient rehabilitation projects (section 6.4)

The total costs of all viable rehabilitation projects usually exceed the available budget. So the projects must be ranked by efficiency criteria in order to spend the money on those projects promising the most positive effects. With regard to the public interest, rehabilitation projects should be ranked on the basis of a comprehensive set of criteria and procedure as well as results should be transparent and easy to understand by political decision makers and the public.

(3) Choosing the right rehabilitation technology (section 6.5)

Beyond the direct costs, the support system will have to take into account a multitude of other factors, which are usually collected by wastewater companies preparing a public tender on a specific rehabilitation project. Although from a financial viewpoint, the wastewater company would choose the lowest bid, it has also to consider external costs that are not charged directly to the wastewater company, such as increased operating costs and travel times for road users. These costs have been elaborated in WP5.

Case study

Each section consists of a theoretic part and examples from the practical application of working package 6 tools. Examples are presented for the Pieschen sub-network of Stadtentwässerung Dresden (SEDD), one of the CARE-S end-users. The sub-network represents 139 km, or approximately 9% of the total network length of SEDD. Construction periods are well known for the network. Just the after World War II decades are rather assumptive, especially the significant peak in 1960 is most probably due to the assignment of pipes for which the exact installation year is unknown (Figure 6.1).

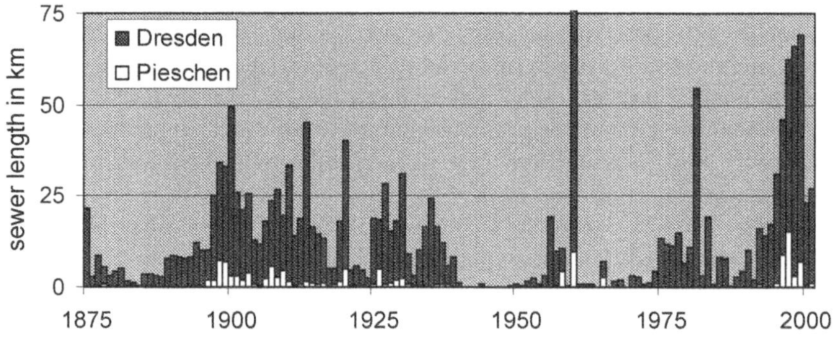

Figure 6.1: Sewer length in km by construction year for Dresden and Dresden-Pieschen

The sub-network is located in the North-Western part of Dresden on the right hand riverside of River Elbe, and it has been chosen with respect to its hydraulic logic (after prior consultation with WP3). The area it covers includes high density urban areas, industrial and commercial areas, a more rural area in the Northern part, and the Wastewater treatment plant (WWTP) (Figure 6.2).

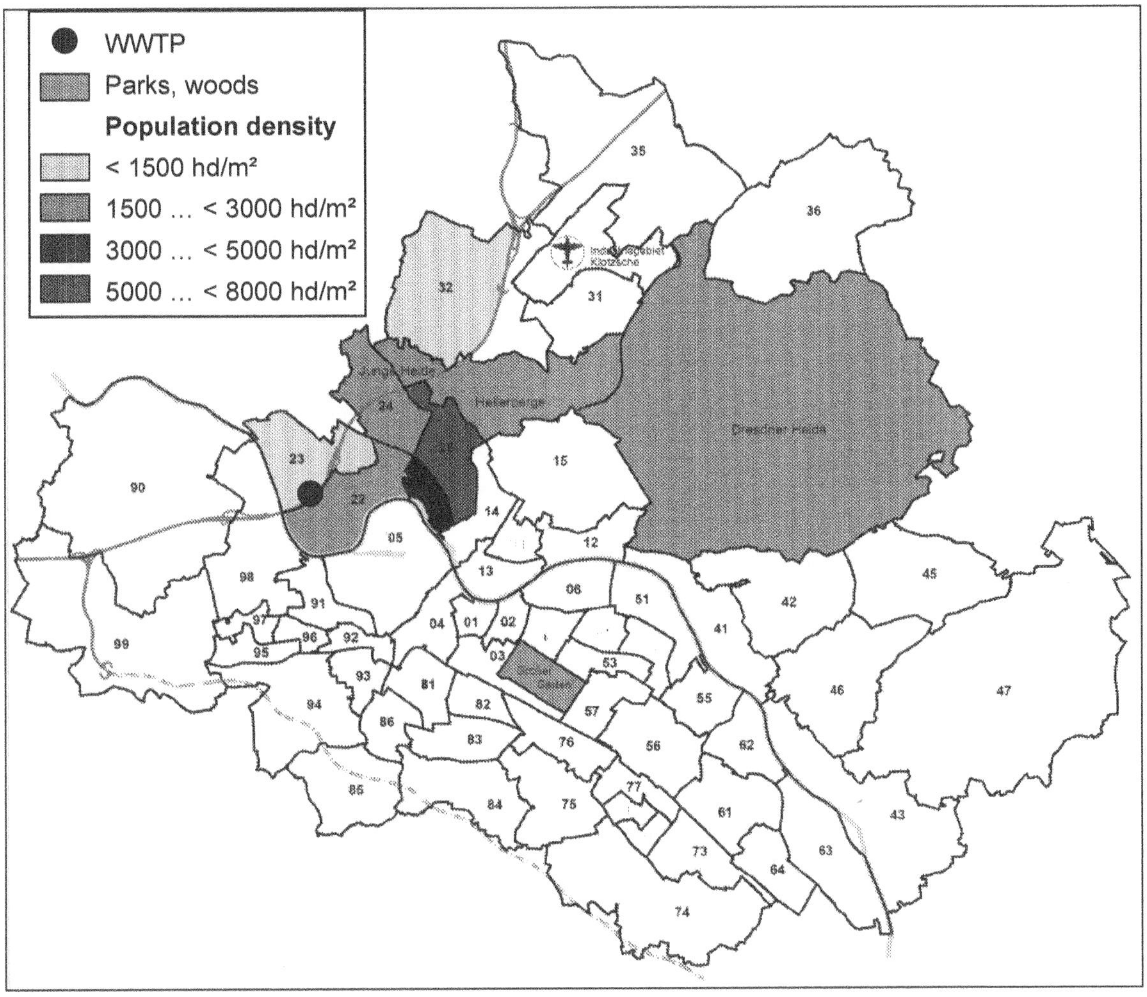

Figure 6.2: Dresden - statistical districts and area covered by the sub-network of the case study; indicated are green spaces, population densities of the area of interest and the location of the WWTP

Figure 6.3: Schematic layout of the Dresden-Pieschen sub-network (left) and condition grades of inspected sewer pipes (right)

6.2 Methodologies of multi-criteria decision support

The objective of a multi-criteria methodology can be to find either a transitive[1] overall final order of a finite set of candidates, or to select a finite number of candidates. The candidates here can be different types of rehabilitation technologies for choosing the best one, or different sewer pipes with serious defects for which a priority list must be found, or different strategies, which must be evaluated and ranked according to their long-term benefits. In any case, the candidates must be compared with respect to a number of criteria C_i, by calculating the criteria values e_{ij} for each candidate P_j (Figure 6.5).

	Rehabilitation candidates			
Criteria	P_1	P_2	P_3	...
C_1	e_{11}	e_{12}	e_{13}	...
C_2	e_{21}	e_{22}	e_{23}	...
C_3	e_{31}	e_{23}	e_{33}	...
...	...	...	...	...

Figure 6.5: Impact matrix

[1] A final order of a set of candidates is said to be transitive if it meets the transitivity condition (e.g. if $P_1 > P_2$ and $P_2 > P_3$, then $P_1 > P_3$). It is said to be linear if there is no cycle of preference involved (i.e. '$P_1 > P_2 > P_3 > P_1$' is cyclic and therefore not linear. See also section 6.5)

Classification of methodologies

Multi-criteria evaluation methods can be classified either with respect to the final result they produce into ranking, classification, and selecting methodologies, or according to the prevailing evaluation principle into substitution methodologies and elimination methodologies.

Substitution methods (scoring models)

A very popular multi-criteria technique, the method of average weighting, belongs to the substitution methods, also referred to as aggregation or scoring models. The overall ranking of candidates is built in three steps:

1. Transformation of all criteria into a uniform scale (normalisation)
2. Distribution of weights, i.e. relative importance or exchange ratios between criteria
3. Aggregation with a utility function for the final ranking order

Usually, aggregation methods are applied, which in most cases use a monetary expression or dimensionless point scales for the normalisation. In general, the normalisation and the definition of the utility function are prepared by an "expert", whereas the only interaction between the decision-maker and the decision procedure is the assignment of weights. Conventional scoring models then apply a simple weighted-sum-rule for obtaining the final result, in terms of priorities, or a ranking order. The problem of these methods is the priority and weighting of the individual criteria, since overrating and underrating of certain aspects is most likely to happen (Vincke 1992). The cost benefit analysis is a sophisticated, but typical substitution method, where all costs and benefits are normalised into the currency scale. The problem of assigning costs to polluted waters, odour, or collapse risks is the well known and often discussed difficulty of this approach (Heinemann 1984, Strassert 1984). An additional aspect and argument against these methods, is the politically accepted incomparability of monetary and environmental or social aspects within decision problems with public relevance (Roy 1985).

Elimination Methods

The principle of elimination methods is to reduce the bandwidth of candidates with thresholds (=arguments) by excluding unsuitable candidates. There is no overall evaluation, but a stepwise threshold setting at particular criteria for eliminating undesired (or less efficient) candidates. One of the advantages of elimination methods is that criteria that are measured in different dimensions need not to be normalized into one common dimension. A very popular, non-formalised version of elimination techniques is the discussion, or - in its more formalised version - the structured interview, for example applied in the course of urban planning competitions. The selection procedure consists of three elementary steps:

1. Setting a threshold value at a particular criterion
2. Observing the consequences at all other criteria
3. Confirming (or rejecting) the elimination threshold

The process is repeated until a break-off limit (e.g. a given budget) is reached.

Within elimination procedures, effect control and reasoning initiate a process of balancing advantages against disadvantages, which leads step-by-step to a common decision. Thus, elimination procedures have rather the character of a decision support system than decision-making, like the substitution methods do.

Multi-criteria pairwise comparison

For a limited number of candidates, a formalized pairwise comparison can be applied with respect to multiple criteria. A typical example for the application of a formalized multi-criteria pairwise comparison is site selection in logistics, e.g. power plant or destructor location. The methodology has its limits in the number of candidates, as the number of pairwise comparisons increases rapidly: for n candidates, the number of comparisons is $n*(n-1)/2$. The result is an overall ranking order of candidates. Like for the elimination methods, criteria need not to be normalised into one common dimension.

Procedures for multi-criteria decision support in CARE-S

The objective of multi-criteria methodologies in CARE-S WP6 is to provide decision-support for three different types of decisions: Developing a long-term rehabilitation strategy, selecting cost-efficient rehabilitation projects, and choosing the best rehabilitation technology. Different procedures of multi-criteria decision support are analysed, that have been applied in the past to decision problems in the field of infrastructure rehabilitation.

The development of long-term strategies is supported by the application of the WP2 structural deterioration model GompitZ. It has been integrated into a framework for the calculation of alternative strategies that can be compared in a systematic way to each other. An example is given for the consideration of predictable performance indicators within the evaluation of rehabilitation strategies. Cost-efficient rehabilitation candidates are selected by the support of an interactive elimination procedure. Here, cost-efficiency is determined by the most positive impacts e_{ij} achieved by all rehabilitation candidates realised within the frame of a given budget. The choice of the best rehabilitation technology for a selected rehabilitation candidate is supported by a formalised balancing and ranking procedure.

6.3 Exploring rehabilitation programmes and strategies

Investments in sewer rehabilitation must be based on inspection and evaluation of sewer conditions with respect to the severity of sewer damage and to environmental risks. This chapter deals with the problems of forecasting the condition of sewers in a network from a small sample of inspected sewers, and the development of an appropriate rehabilitation strategy for meeting certain objectives. Key issues of strategic or long-term rehabilitation planning of a sewer network are:

- Forecasting the future condition of sewer pipes by deterioration models
- Determination of deterioration rates and the associated rehabilitation needs for
  (a) maintaining the actual condition of the network in the long term
  (b) reaching an improved level of service within given time horizon
- Forecasting performance indicators of the network for different strategies (scenarios)
- Evaluation of different scenarios, and choosing the best rehabilitation strategy

Within CARE-S, condition forecasts are performed using the CARE-S WP2 GompitZ model (Le Gat 2004). Transition functions from one into the next poorer condition class are empirically derived from a sample of inspected sewer pipes. The transition probabilities are used to forecast the future condition of sewers. By the same procedure, transition probabilities have been calibrated for sub-samples of different types of sewers. With these probability functions, the most probable date of entering a critical condition class can be forecast from pipe characteristics, such as material, period of construction, location, use for waste and/or storm water, profile, diameter and so on.

Data set

Obviously, it is inefficient to monitor permanently the condition of the complete sewer network. In order to prevent sewers from collapse, choke, wastewater overflow or exfiltration and groundwater infiltration, the state of the sewers must be inspected, not necessarily at regular intervals but in due time before serious damage occurs. A representative sample of the network pipes should be inspected for getting an idea of the condition of the overall network, and in order to develop a strategy against its deterioration. Sewers are deteriorating slower or faster under specific local circumstances. So, their condition is not determined by age alone. There are variations in material, load, stress, wastewater and subsoil characteristics which have to be considered as factors influencing the process of sewer deterioration. Procedures for how to create a representative data set from limited inspection data are discussed for example in Sjøvold et al (2005) and Baur and Herz (2001). A detailed statistical description on how to select sewer pipes for a selective inspection plan and on the subsequent estimation error of the condition of the total network is discussed in BMB+F (2004).

The efficiency of sewer inspection is greatly improved if failure prone sewers were inspected more frequently and others at larger intervals. In other words: The same risk level could be guaranteed with a smaller amount of inspection; or, with the same inspection cost, the risk of failures could be reduced if local knowledge on specific sewer deterioration is used systematically for inspection planning. A general rule, how to stratify the sample, cannot be given. However, in most cases the main items for distinction will be material (concrete, clay, brick, plastic,...) and construction period. In Dresden, for instance the following periods have been chosen: before 1900, 1900-1940, 1940–1989, after 1990. The user of the GompitZ model is free to define the types of sewer pipes. The stratification variables investigated for the application of the GompitZ model in Dresden-Pieschen are material and construction periods (Figure 6.6, Table 6.1).

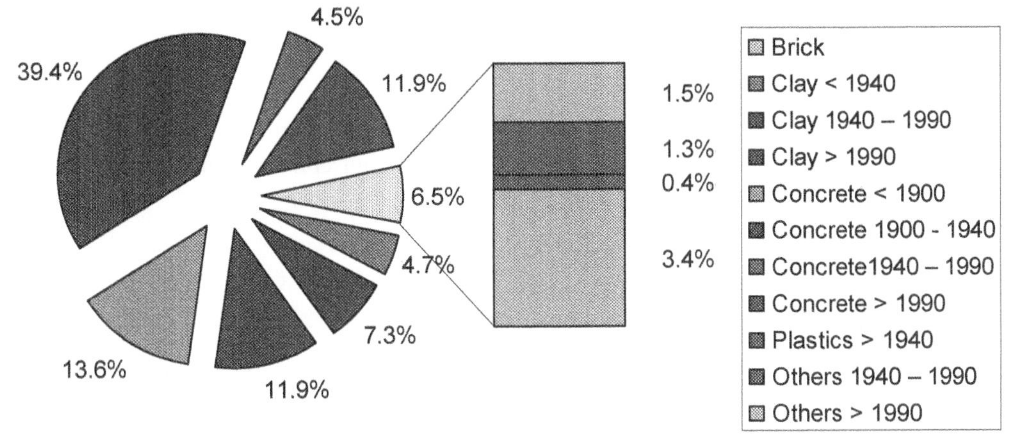

Figure 6.6: Sewer pipes by material and construction period in Dresden-Pieschen

Plastic pipes are mainly made of PVC. The group of *other* materials consists mainly of asbestos cement pipes installed in the 1960ies (with no inspection results available), and ductile iron pipes installed in the 1990ies.

Table 6.1: Percentage of pipes by material and construction period and CCTV inspection ratio

Material	Construction period	Length in km	Length [%]	inspected
Brick		2.1	1.6 %	24.7 %
Clay	< 1940	6.6	4.7 %	21.6 %
	1940 – 1990	10.1	7.2 %	50.8 %
	> 1990	16.6	11.9 %	51.3 %
Concrete	< 1900	18.9	13.6 %	42.6 %
	1900 - 1940	54.8	39.4 %	58.4 %
	1940 – 1990	6.3	4.6 %	20.4 %
	> 1990	16.6	12.0 %	51.3 %
Plastics	> 1940	1.8	1.3 %	33.6 %
Others	1940 – 1990	0.5	0.4 %	0 %
	> 1990	4.7	3.4 %	81.5 %
Total	**1828 - 2001**	**139.0**		**51.7 %**

Condition forecast with GompitZ

A detailed description of the non-homogeneous Markov-chain approach in CARE-S, GompitZ, is given in chapter 2 of this report. The principle idea behind Markov-chains is to determine the probabilities of the transition from one condition state into the next . For example, the probability to be in condition grade 3 at a certain time is a function of the transition probability from condition grade 2 to 3, the transition probability from condition grade 3 to 4, and the probability to remain in condition grade 3. The probabilities are calculated by the statistical analysis of asset data, using information either for stratification of the sample or as explanatory variables. The phrase "non homogeneous" here means that transition probabilities are time dependant. The condition forecast is calculated by using a Weibull distribution (Gompertz). The model can use any kind of variables for the analysis. The minimum information required is data on:

- PipeID
- Construction year
- Inspection year
- Length
- Condition grade

Any number of variables can be added, as long as the information is complete for all pipes. Then, the variables explanatory vector is calibrated with empirical data from the random inspection sample. Explanatory variables available from the case study, which have been used for calibration, are:

- dimension (height in mm)
- sewer type (fowl, storm, combined)
- depth (average in m)
- under groundwater table (yes/no)

Finally, for each pipe, year by year the probabilities of staying in the different condition grades are calculated, cumulated and multiplied by the total network length:

PipeID;year;p(1);p(2);p(3);p(4);p(5);p(6)
Pipe0020;2005;0.485272;0.362755;0.110884;0.033396;0.007693
Pipe0020;2006;0.449517;0.383841;0.121303;0.036835;0.008504
Pipe0020;2007;0.413550;0.404113;0.132389;0.040560;0.009387
Pipe0020;2008;0.377670;0.423249;0.144144;0.044591;0.010347

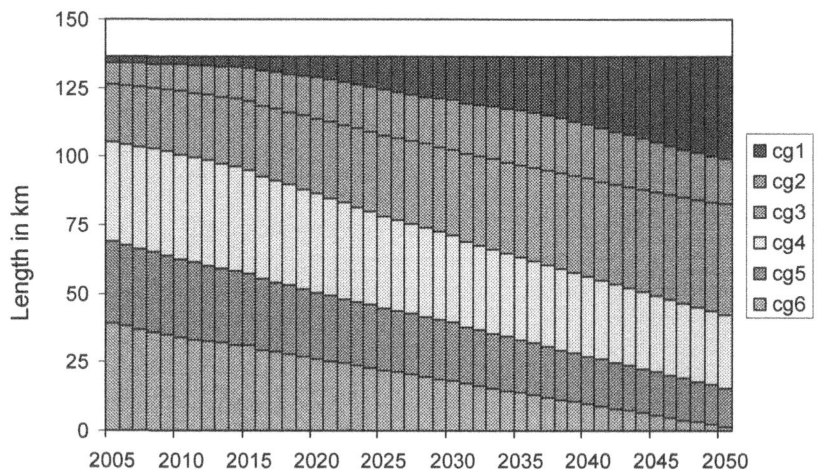

Figure 6.7: Forecast network condition

Options for future rehabilitation strategies

The condition forecast itself assumes a "zero case" in which no rehabilitation is carried out in the future, and the network deteriorating is "undisturbed". Three types of intervention are defined as potential rehabilitation strategies:

1. Length driven strategy
2. Budget driven strategy
3. Condition target strategy

The first two types of strategies assume a given length or budget for a certain planning period. Within the third strategy, for a given period, e.g. 10 years, a certain deficiency (e.g. pipes in the worst condition grade), shall be reduced, in order to achieve a certain service level, that shall be kept in the long-term.

Example for the consideration of predictable PI: Infiltration

In a joint study of WP3 and WP6 (Schulz/Baur/Krebs 2004), infiltration volumes and infiltration likelihood has been cross analysed with the assignment to condition grades of pipe types. In addition, the subsequent potential reduction of infiltration volumes has been considered within a hydraulic model for calculating the effects on CSO performance and WWTP inflow. Large diameter concrete pipes show a exponential correlation between their degree of deterioration, respectively condition grade, infiltration rate, and the pipe length affected by groundwater. How far this correlation is caused by the assignment of pipes to condition grades 3, 2 and 1 due to observed infiltration will be subject to further data analysis (Rutsch/Uibrig 2003). In addition, infiltration is also influenced by other factors, for instance the additional load on these pipes due to groundwater pressure, which is considered in the leakage approach applied by Karpf/Krebs (2004), changing loads due to variations in the groundwater level, and others.

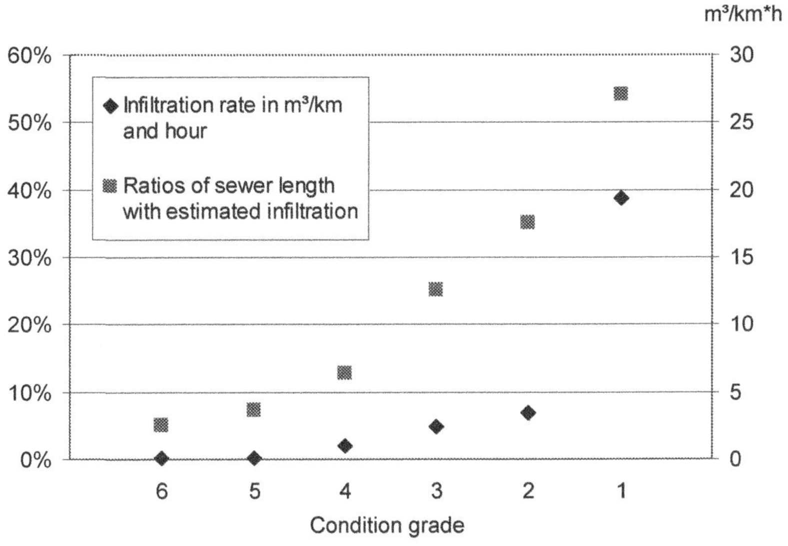

Figure 6.8: Infiltration length and infiltration volumes against condition grades for large diameter (> 800mm) concrete pipes (Schulz/Baur/Krebs 2004)

In general, the application of a deterioration model using condition grades could be suitable for the calculation of future infiltration rates of parts of the network. Thus, rehabilitation strategies could focus on critical areas in the network, where infiltration volumes should be reduced. The assumption of reducing infiltration rates by a certain rehabilitation length is limited, because the water will probably find other ways to enter the system if leaky sections are tightened, and in addition, that the number of pipes prone to groundwater is limited. On the other hand, the deterioration model could simulate the increase of infiltration rates in ranges into the system due to its higher likelihood of receiving groundwater for particular condition grades. According to the ageing forecast, the ratio of large diameter concrete pipes deteriorated to condition grade 3 or worse, and thus prone to higher infiltration, would reach approximately 80% of the total length of the sub-sample within the next three decades (Figure 6.7).

With respect to their infiltrating lengths, large diameter concrete pipes of condition grades 2 and 3, are in the average 2.7 times more likely to exhibit infiltration than those of condition grades 4 and 5. Specific infiltration rates into the pipes of condition grades 2 and 3 are 2.1 times higher than in the respective pipes of condition grades 4 and 5. So, as an input parameter for the integrated model, a factor for the infiltration rate of the deteriorated network could be derived for the comparison of rehabilitation with the "do-nothing" option. This factor may be in the range of 1.4 to 1.8 times of the actual infiltration rates.

Beside other criteria, the performance of the system has been assessed for CSOs by the calculation of the number of overflow spills, the overflow volume, and the duration of overflows, as a function of the infiltration rate. The parameter has been varied between 0%, 40%, and 80% additional volume compared to the dry weather flow.

Figure 6.9 shows the simulated influence of infiltration on the overflow volume and duration. The various criteria to assess the effects of the CSO do not always coincide in describing the performance of the overflow structure and are affected differently by the increased input of extraneous water.

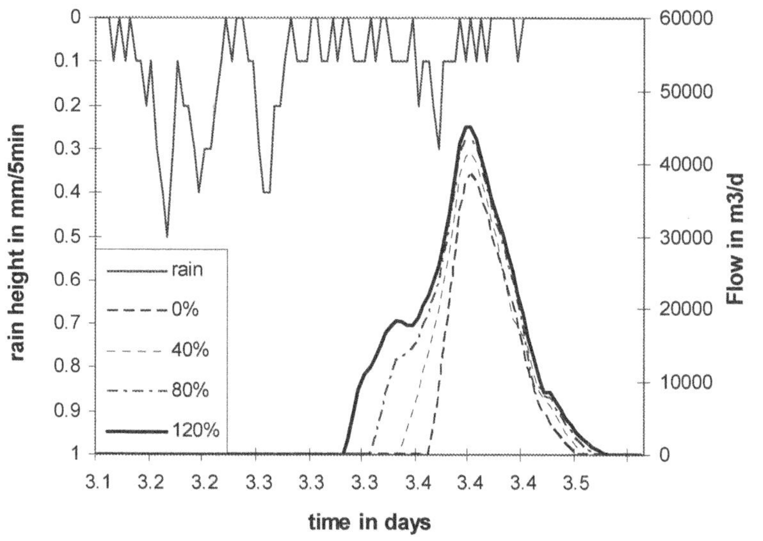

Figure 6.9: CSO flow as a function of the extraneous water rate (Schulz/Baur/Krebs 2004)

Figures 6.10 to 6.12 show three criteria applied to a virtual test catchment with three overflow structures (Structure 1, Structure 2 and Structure 3) with retention tanks. The acute impact of CSOs can be expressed by analysing the minimum oxygen concentration, the maximum ammonium concentration, and the duration of the violation of the threshold value. In this analysis thresholds have been set to

- minimum oxygen concentration = 5 mg/l
- maximum ammonium concentration = 4 mg/l.

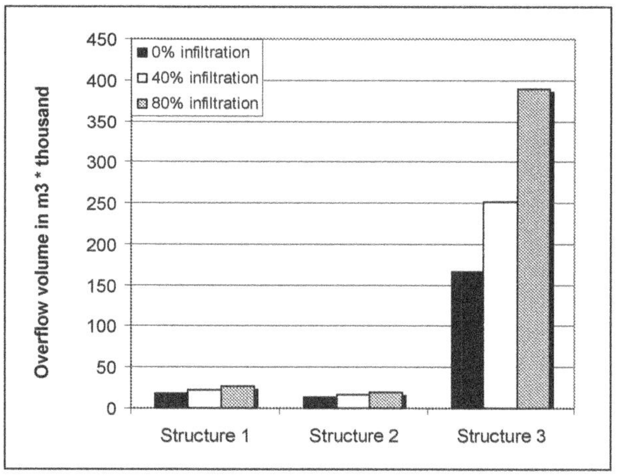

Figure 6.10: Overflow volume

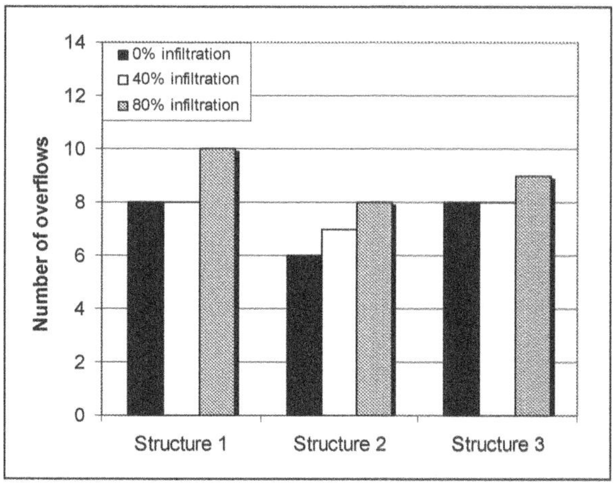

Figure 6.11: Number of CSO spills

The effect of infiltration volumes on the maximum concentration of NH_4^+ and on the oxygen minimum is not very distinct compared to the absolute values. However, the duration of the critical concentrations is heavily affected. This is an important factor, since the evaluation of the acute impacts is based on two parameters, concentration and duration.

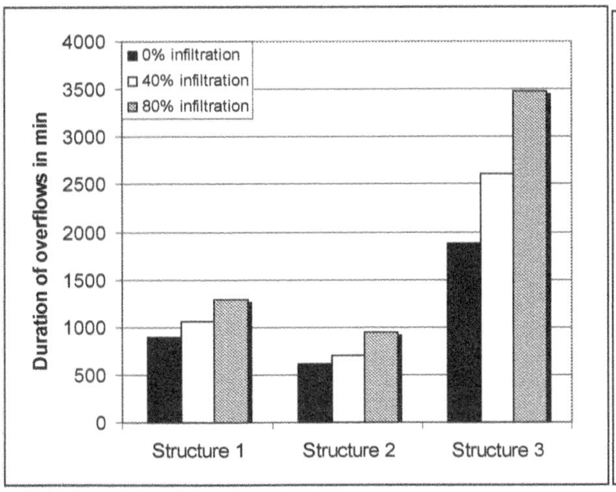

Figure 6.12: Duration of CSO spills

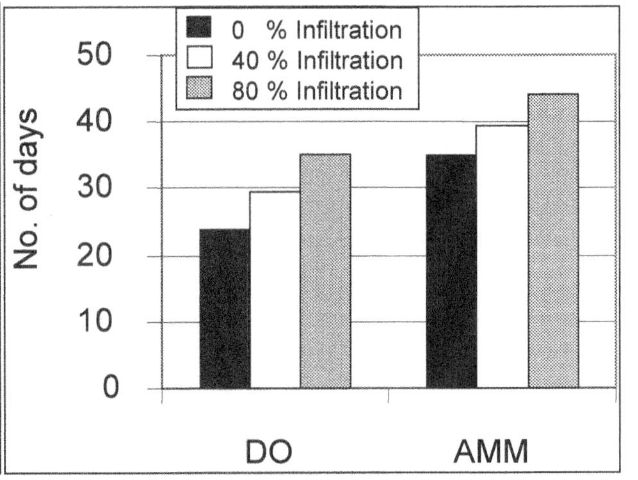

Figure 6.13: No. of days with critical oxygen minimum and Ammonium concentration

The study shows that the effects of an increase in the infiltration rate into the system can be estimated by the integrated model. Results then can be related to the costs of avoiding these impacts. In that case, the network deterioration model can be used for the calculation of rehabilitation needs in terms of an annual budget, required for maintaining the present network condition in terms of the percentages of condition grades in the network for certain years in the future.

The objective of the study (Schulz/Baur/Krebs 2004) was the estimation of potential benefits of sewer pipe rehabilitation for the performance of the drainage system and the wastewater treatment plant (WWTP) as well as for the receiving water quality. The relation of sewer system status and the infiltration rate was assessed based on the statistical analysis of 470 km of CCTV inspected sewers of SEDD, Dresden. Critical pipe types have been identified and it was shown that the potential reduction of infiltration volumes and the consequent performance improvements of the urban wastewater system can be simulated as a function of rehabilitation activities in the network. The integrated model was applied to a virtual system with input from a real sewer network.

Infiltration rates will only decrease significantly if large sewers or large parts of a network are rehabilitated. After rehabilitation the groundwater level may rise or the groundwater will search new pathways to infiltrate into another location (pipe) of the sewer system. If a procedure is available to estimate the infiltration rate before and after rehabilitation, the integrated simulation can deliver indicators to relate the rehabilitation costs to the improvement of the quality of receiving waters. In Germany, concentrations of WWTP effluents are used to monitor the performance. If a violation of the threshold concentration is associated with costs, the lower pollution concentrations in catchments with a higher infiltration rate could even be an advantage, and thus it could be an argument *against* rehabilitation measures although the operational costs in the WWTP might increase and higher pollution loads are discharged. Therefore, the definition of criteria used in the assessment of infiltration effects is a very important issue that should be handled carefully.

The application of an integrated model for the evaluation of rehabilitation including infiltration effects is only useful if sufficient data is available in all compartments of the system and if a method can be applied to estimate the infiltration as a function of the condition state of the sewer system. Both requirements are not fulfilled completely until now. Within the study, so far, infiltration volumes were subject to parametric variations in the integrated model, and the effects of different infiltration rates on the urban water system were simulated, e.g. in terms of CSO volumes and frequencies. By the deterioration model the length of pipes that will enter worse condition grades within a defined forecast horizon can be calculated. Even if there is not a causal connection between condition grade and infiltration volumes, such a factor could be a suitable input parameter for the integrated model and for estimating the efficiency of a strategy.

6.4 Selecting cost-efficient rehabilitation projects

In general, the total cost of all viable rehabilitation projects exceed the available budget, thus not all sewers in critical condition can be rehabilitated immediately, and it becomes necessary to select those rehabilitation projects promising highest cost efficiency.

Criteria for the selection of rehabilitation projects in a sewer network are the different classification results and priorities with respect to structural and hydraulic aspects, environmental aspects and the associated direct and indirect costs. External effects are considered by the risk associated with the deficiency of the sewers, for example the potential damage due to frequent flooding or the potential groundwater pollution due to an exfiltrating sewer. Supported by a geographical presentation of the network (Figure 6.14), the candidates for rehabilitation are selected, step-by-step, in an interactive elimination process. Within the procedure which was successfully applied to the rehabilitation planning of water distribution networks, those candidates are eliminated which are not fulfilling specified thresholds. The remaining candidates have a bundle of attributes favourable for rehabilitation (Baur 2003).

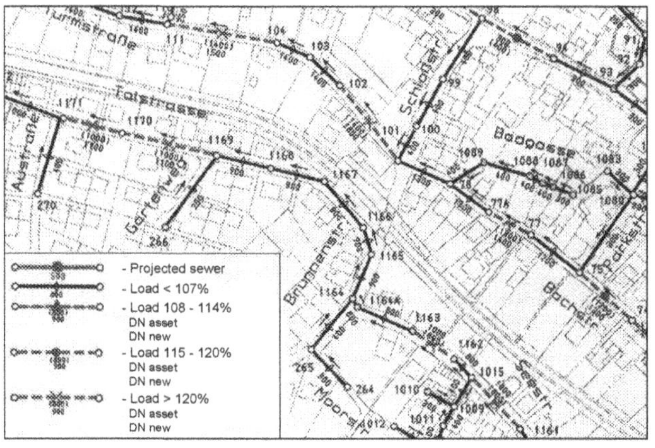

Figure 6.14: Rehabilitation priorities in a network due to hydraulic loads (MUV-BW 2000)

The interactive elimination process has been developed for the rehabilitation of infrastructure networks, and modified for the development of sewer network rehabilitation plans in CARE-S. Alternatively, the concordance analysis developed for CARE-W (Le Gauffre et al. 2002b), has been subject to investigations within CARE-S.

Interactive Elimination Procedure (IEP)

The IEP is a procedure that frequently exists in a non-formalised variant in many fields of technical and political decision making where public is involved. A typical example is urban planning, where the candidates are elaborated within a competition, and a jury is setting the thresholds for the elimination of candidates that are not good enough. Then, some candidates may be better in design, others in functionality, and a third group is cheaper. It's up to the jury to set the threshold for costs, and to check what functionality and design is still affordable. If they agree, the threshold is accepted, and the set of candidates is eliminated, until a final solution, or set of drafts for a more detailed planning, is selected. The objective of the engineer is the appropriate presentation of information and criteria, in order to illustrate the required trade-offs of a decision. This decision methodology can be best characterised as a feasibility and sensitivity analysis by multiple criteria. It aims to control the impact of the performance of other criteria, if for one criterion an elimination threshold is set for rehabilitation projects that are "not so good" (Gero 1982, Hochstrate 1986, Baur 2003).

In the schematic outline of Figure 6.15 the procedure is demonstrated for 4 criteria: a first elimination threshold (1) is set for criterion C_1. This leads to the elimination of a certain number of projects (grey dashed areas). The elimination of these candidates does not consequently reduce the performance range of the other criteria as desired: for example, the elimination of pipes with a higher blockage risk will not necessarily lead to the elimination of those pipes that probably cause major odour problems in the network. Therefore, elimination thresholds must be accepted or not (or modified (2)) after an impact control of the respective elimination step. Once an elimination step is confirmed, the next threshold is set at another criterion (3). By this procedure step by step those candidates for rehabilitation will remain that have a bundle of impacts favourable for rehabilitation. The process comes to an end, when the break-off limit is reached, which could be, for example, a budget limit or a certain rehabilitation rate.

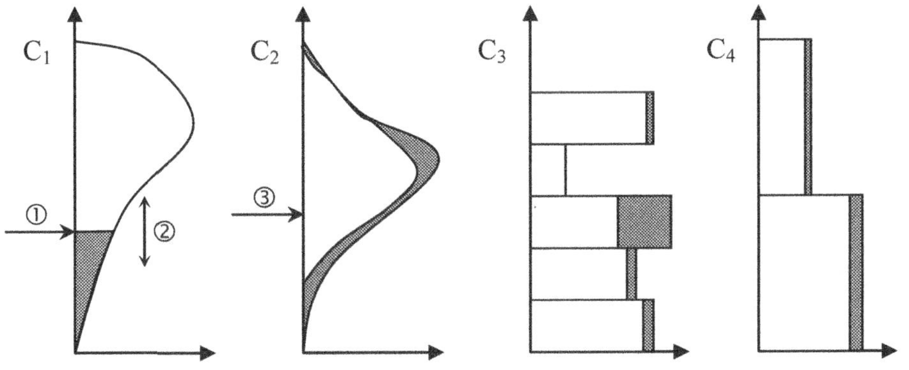

Figure 6.15: Schematic outline of the interactive elimination with effect control (Baur 2003)

The order, in which the criteria are chosen for the elimination steps, is free. Mathematically, the result will not be influenced by the order of application of decision criteria, due to the conjunctive nature of the thresholds (mathematically, the elimination conditions are joined by a logical "OR") (Zimmermann and Gutsche 1990). However, different preferences might be expressed by the decision maker, in the case of a changing order of decision criteria. In practice, the elimination will start with "hard" criteria, for which the decision is not so difficult, and where just minor trade-offs must be accepted, for example for younger pipes of new materials with condition grades better than the network mean. As the decision process advances, more and more "softer" decision criteria will be applied: e.g. the presence of sensitive customers or complaints. This way, the decision maker's understanding of the consequences of his decisions and the possible trade-offs will be improved step-by-step (Hochstrate 1986).

For supporting a reasonable threshold setting, different presentation tools are applied and developed for use in the elimination process: tables, charts, histograms and cumulative curves are combined with a map for the spatial presentation of the selected projects (Baur 2003). This way, the remaining projects and the eliminated projects can be presented, and a permanent control of the elimination process is possible.

Formalisation of the procedure

The interactive elimination procedure for the selection of efficient pipeline rehabilitation projects is given a formal structure that is outlined in Figure 6.13. The procedure starts with the preparation and organisation of existing information and the development of the decision matrix. In the decision matrix, potential rehabilitation projects must be clearly identified entities that are associated with their performance at different criteria. Then, a break-off limit must be defined. It is advisable to define the reference solution at certain decision criteria, from which an individual effect control is carried out, with the cumulative curves of the respective criteria. The reference values can be modified during the elimination procedure without any direct influence on the final solution.

A first elimination threshold is set for an attribute that allows without many difficulties the elimination of the projects. Hence, the elimination threshold should not be too rigid. The affected pipes are marked and, temporarily, eliminated from the final solution. In the case of projects that will be definitely selected for the rehabilitation plan, for example due to co-ordination with road works, these projects can be labelled as "finally selected".

After the temporary elimination of pipes by the first elimination criterion, the effects of this elimination must be controlled. An overview of the selected pipes gives the map, the changes in the criteria performances can be observed by the histograms, charts or cumulative curves. If any projects exceed one of the reference values, the elimination must be confirmed individually. If the reductions of the criteria performances are accepted, and if no eliminated project exceeds for any criterion its reference value, the elimination is confirmed and the projects are finally eliminated. Now the decision maker can decide, whether to increase the threshold for the first elimination criterion, or to choose another criterion, for which a new threshold is set. If the restrictions and trade-offs after the elimination can not be accepted, the threshold must be modified.

Then the next elimination criterion is chosen, and the procedure continues until the break-off limit is reached. When the procedure comes to its end, not necessarily all criteria have been used for elimination. In fact, there will be some criteria that preferably might be used just as "control-criteria" by the decision maker, whereas others are mainly used for the active threshold setting.

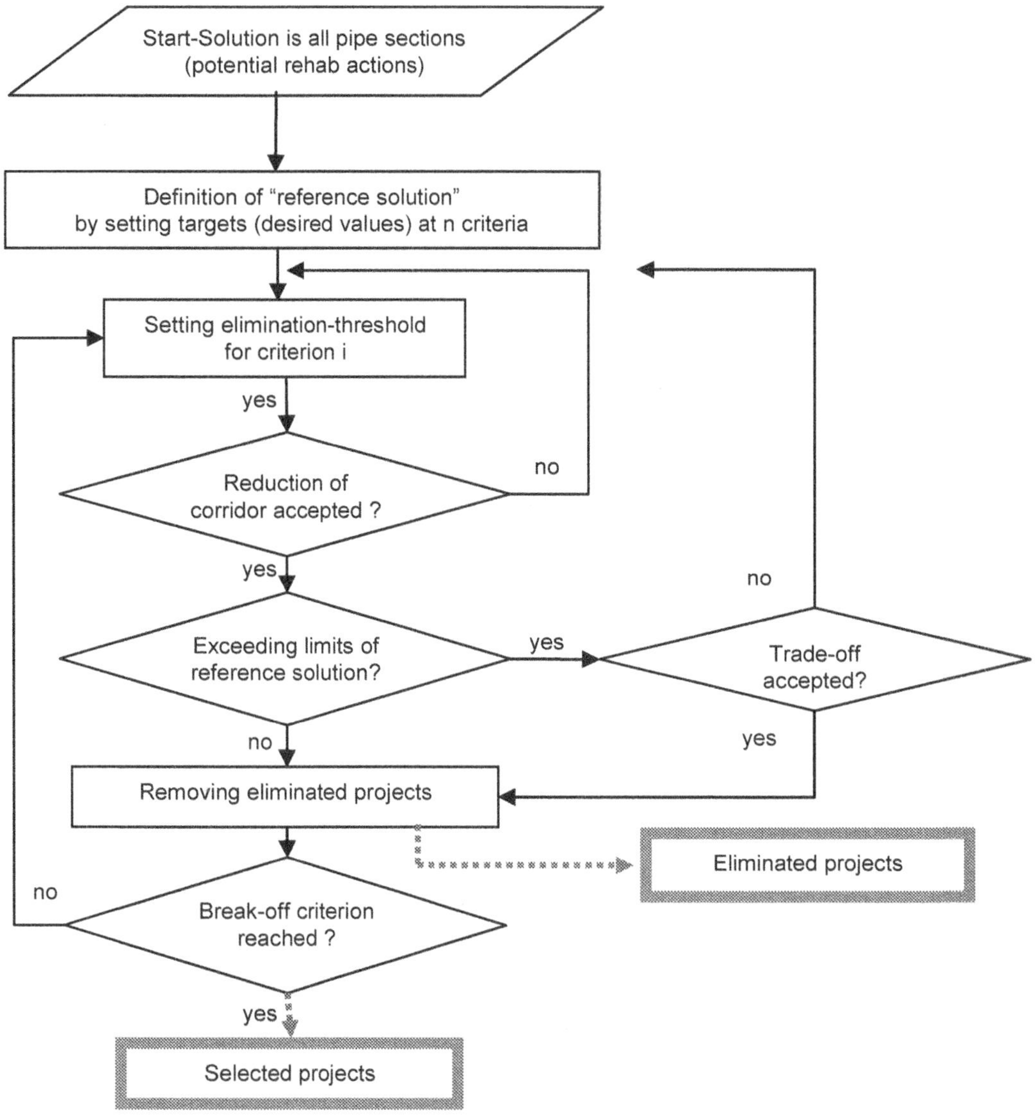

Figure 6.16: Interactive elimination procedure, flow chart (Kropp and Baur 2002)

If in the decision maker's opinion no further trade-offs could be accepted before the break-off limit is reached, new alternatives must be developed (for example by the design of operational changes in the sewer system), the constraints must be changed (for example by increasing the budget), or the information base must be improved, in order to have a better knowledge base for setting priorities between the projects. Each elimination step is documented in a protocol with the thresholds, the modifications and the confirmation status (yes/no) set. Within the protocol the argumentation for the acceptance of trade-offs can be reported, and the voting behaviour in majority decisions is documented, as well. The result of the procedure is presented in lists and maps, and it includes the minutes of the decision process. The procedure will not necessarily converge to results in the sense of a mathematical optimum, but it will lead to a robust solution in the sense of giving a better understanding of the problem (with respect to data availability and requirements) and it will stimulate a learning process, which should be the objective of any method of decision-support.

Application

Decision criteria for the setting of priorities consider the actual hydraulic (functional) and structural condition of the pipes and the potential impacts in the case of malfunction. Within CARE-S, decision criteria are the condition grade, and those defined by WP3 (Hellmud tool) and WP5 (socio economic evaluation). The tool for priority setting does not require a complete list of all criteria.

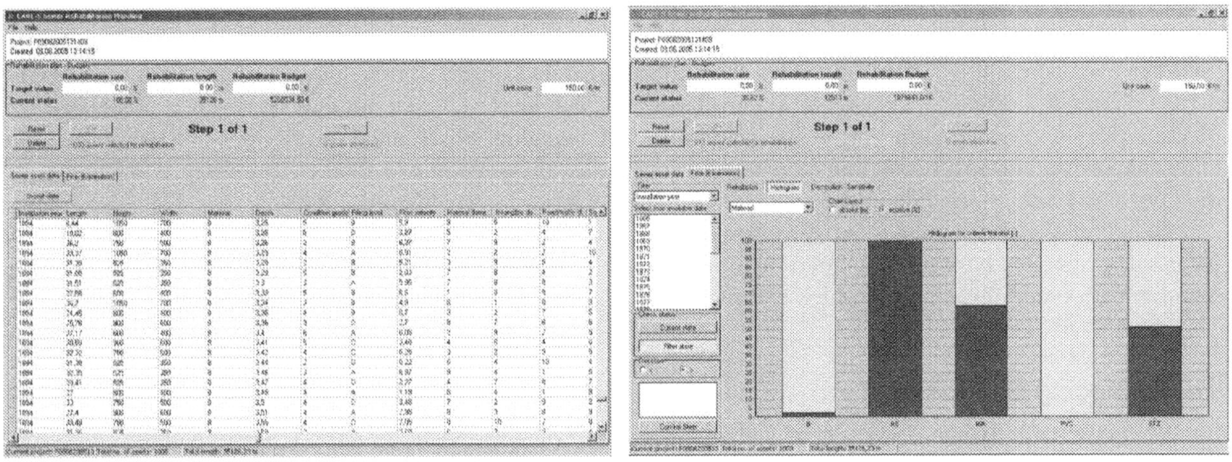

Figure 6.17: Interactive elimination procedure, application screens

6.5 Choosing the right rehabilitation technology

The formalised balancing and ranking procedure (BRP) presented by Strassert (1995) and Köhl (1998) is a new variant within the field of semi-formalised multi-criteria decision support methodologies. "Basic features of the approach are the pairwise comparisons of options, mixed scales and the so-called balancing principle, i.e. the balancing of vectors of advantages and disadvantages" (Strassert 2000: 1). For any pair of options the relative advantages and disadvantages are balanced and, simultaneously, the different importance of the score is taken into account. Strassert (2000) gives a detailed derivation and description of the methodology. It has been applied in CARE-W to the evaluation of rehabilitation strategies (Herz et al. 2003), and to the choice of the best rehabilitation technology for sewer pipes (Schmidt 2002, Plenker 2003).

Preliminary order of candidates

The procedure starts with the impact matrix (Figure 6.5). Each criterion C_i yields an individual order of candidates P_j by means of e_{ij}. For $e_{11} > e_{12} > e_{13}$, the ranking order of criterion C_1 is P_1, P_2, P_3. In general, the ranking order of candidates P_j will differ for all criteria C_i.

Now suppose that the criteria values e_{ij} show that, for C_1, $P_1 > P_2 > P_3$. Then we can write:

C_1: < P_1, P_2, P_3 >, and similarly for C_2 and C_3, for example

C_2: < P_3, P_2, P_1 >,

C_3: < P_3, P_1, P_2 >,

In an "outranking matrix" (Figure 6.18a), the number of pros and cons are counted for each candidate. The matrix in Figure 6.18a shows that candidate P_1 is placed twice (i.e. for two criteria) before candidate P_2, once before P_3, and vice versa P_2 is placed once before candidate P_1, and P_3 is placed twice before candidate P_1 (The table must be read in rows).

	P_1	P_2	P_3
P_1	0	2	1
P_2	1	0	1
P_3	2	2	0

Figure 6.18a: Outranking matrix

	P_3	P_1	P_2
P_3	*	2	2
P_1	1	*	2
P_2	1	1	*

Figure 6.18b: Triangular outranking matrix

When the "outranking matrix" is set up, the options must be re-ordered to achieve a triangular matrix (Figure 6.18b). Triangularisation is the systematic re-ordering of candidates such that out of a set of p = j! orders, the sum of the values above the main diagonal is maximised. A situation where only zero values are below the main diagonal corresponds to a strong transitive overall final order of candidates, or total order structure. Normally, this total order structure of candidates does not exist initially (Vincke 1992, Strassert 2000).

The outranking matrix relates to the majority rule of counting votes (1 criterion → 1 vote). In the context of the balancing principle, the assumption of a majority rule must be given up. The outranking matrix is used instead as a preliminary order of candidates which is then subject to a screening and balancing process.

The balancing process

Firstly, a new table is introduced. In an "advantages-disadvantages table" the criteria C_i are combined with the result of the pair wise comparison of candidates P_j/P_k (Figure 6.19). The column headings contain all possible comparisons in pairs. For n candidates the number of comparisons is $z = n*(n-1)/2$. In our case ($n = 3$), $z = 3$.

	P_1/P_2	P_1/P_3	P_2/P_3
C_1	A_{P1C1}	A_{P1C1}	A_{P2C1}
C_2	D_{P1C2}	D_{P1C2}	A_{P2C2}
C_3	A_{P1C3}	D_{P1C3}	D_{P2C3}
ΣA_j	2	1	2
ΣD_j	1	2	1

Figure 6.19: Advantages-disadvantages table

For each comparison in pairs, the scores of e_{ij} must be compared. The comparisons can be made independently from the criteria's scales. They refer to quantities, rankings or frequencies (cardinal, ordinal, nominal scale). At the bottom of the table two rows show the sum of advantages ΣA_j and the sum of disadvantages ΣD_j by comparison.

In the advantages-disadvantages table, each column represents a separate binary decision problem: in the first column, the comparison P_1/P_2 of the two candidates P_1 and P_2, the question is whether the two advantages [${}_{1/2}A_1$, ${}_{1/2}A_3$] together are strictly superior (or not) to the disadvantage ${}_{1/2}D_2$. For "Yes", P_1 is strictly superior to P_2. If respectively the answer is "No", then P_2 is strictly superior to P_1. If the answer to the question: Are [A_{P1C1}, A_{P1C3}] strictly superior to D_{P1C2} is "Yes", the disadvantage ${}_{1/2}D_2$ loses its importance.

Thus, each comparison result is rated according to a Boolean code (1,0), and the results of all comparisons can be drawn to a compatibility matrix C (Figure 6.20a), and if the compatibility matrix is reordered, to the triangular compatibility matrix C^T (Figure 6.20b).

	P_1	P_2	P_3
P_1	0	1	1
P_2	0	0	0
P_3	0	1	0

Figure 6.20a: Compatibility matrix C

	P_1	P_3	P_2
P_1	0	1	1
P_3	0	0	1
P_2	0	0	0

Figure 6.20b: Triangular compatibility matrix C^T

The triangular compatibility matrix C^T is representing a "strict total order" with a binary relation, which is asymmetric, complete and transitive (Vincke 1992). With this matrix, the overall final order of candidates looked for is obtained, i.e. $<P_1, P_3, P_2>$ (Strassert 2000).

Operating comparisons in pairs

In practice, the number of candidates, e.g. rehabilitation technologies, to be compared can easily be 5, 6 or more. Then the number of comparisons in pairs z is 10, 15, or more, respectively. Irrespective the number of criteria, the determination of the order relations could become a clumsy job. Not necessarily all possible comparisons in pairs and corresponding balancing problems must be carried out. For 4 candidates, the number of comparisons would be 10. The number of possible orders of candidates is p = n! = 1*2*3*4 = 24. These orders are:

P_1, P_2, P_3, P_4 — P_2, P_1, P_3, P_4 — P_3, P_1, P_2, P_4 — P_4, P_1, P_2, P_3
P_1, P_2, P_4, P_3 — P_2, P_1, P_4, P_3 — P_3, P_1, P_4, P_2 — P_4, P_1, P_3, P_2
P_1, P_3, P_2, P_4 — P_2, P_3, P_1, P_4 — P_3, P_2, P_1, P_4 — P_4, P_2, P_1, P_1
P_1, P_3, P_4, P_2 — P_2, P_3, P_4, P_1 — P_3, P_2, P_4, P_1 — P_4, P_2, P_3, P_3
P_1, P_4, P_2, P_3 — P_2, P_4, P_1, P_3 — P_3, P_4, P_1, P_2 — P_4, P_3, P_1, P_2
P_1, P_4, P_3, P_2 — P_2, P_4, P_3, P_1 — P_3, P_4, P_2, P_1 — P_4, P_3, P_2, P_1

If n - 1 comparisons in pairs are already executed, and if all candidates are considered in the balancing process, the remaining comparisons are given implicitly. For example, if n - 1 = 3 comparisons yield the three orders $<P_1, P_2>$, $<P_1, P_4>$ and $<P_3, P_2>$, then from the 24 orders above, in a first step 12 are eliminated (all orders where P_1 is before P_2), in a second step 8, and in a third step 3 orders are eliminated, respectively. Hence, only one single order, $<P_1, P_4, P_3, P_2>$, remains, that is the final order of candidates.

Intransitivities occur when the balancing result of three comparisons would give rankings of the form: $<P_1, P_2>$, $<P_2, P_3>$ and $<P_3, P_1>$, i.e. the circular relation $P_1 > P_2 > P_3 > P_1$. Strassert (1999) gives an analytic proof that the number of comparisons in pairs could be reduced down to a minimum of n–1 comparisons, and a methodology how to avoid intransitivities in the final order. For the pre-elimination of candidates, knockout criteria are introduced (e.g. applicability conditions).

Solving balancing problems

The solution of the balancing problem is the comparison of two candidates with respect to the set of advantages against the set of disadvantages, presented in the "advantages-disadvantages table". That is the answer to the question: Are the advantages $[_{1/2}A_1, {}_{1/2}A_3]$ strictly superior to the disadvantage $_{1/2}D_2$? The number of criteria in each comparison with an equal score is not decisive. A dominance check looks for comparisons, where we have only advantages but no disadvantages. Rankings including the inverse of such a dominance order can be removed from the set of possible rankings.

Application and workflow model

The formalized balancing and ranking procedure (BRP) is implemented into the CARE-S tool box. A list of projects for rehabilitation and their description is provided by the multi-criteria tool for the selection and prioritisation of rehabilitation candidates (see section 6.4). The rehabilitation technology catalogue is defined in the data base developed within chapter 4 of this report.

After creating a "rehabilitation project", some general information such as project name and description must be given. Subsequently, it is necessary to import the rehabilitation technology catalogue, the project list, and to define constraints for the comparison. The knockout-check is performed then. Automatically those rehabilitation technologies are eliminated, which do not fulfil the required applicability and performance conditions. In addition, rehabilitation technologies can be eliminated individually by the user. The next step is the automatically performed dominance check.

The tool generates now a revised impact matrix consisting of the remaining rehabilitation technologies and all criteria including the according pre-suggested rankings. These rankings can be revised by the user. Then, the weighing and ranking process starts with the first comparison in pair. For this purpose an advantage-disadvantage table is created for each comparison. Here, the user can arrange the order of criteria and determine the "winner". The user must confirm his decision. This process will be repeated until the final ranking has been fixed. During the process, *CARE-S - SRT* checks for intransitivities. Finished decision runs cannot be modified any more, but analyzed. The principal workflow structure is shown in Figure 6.17.

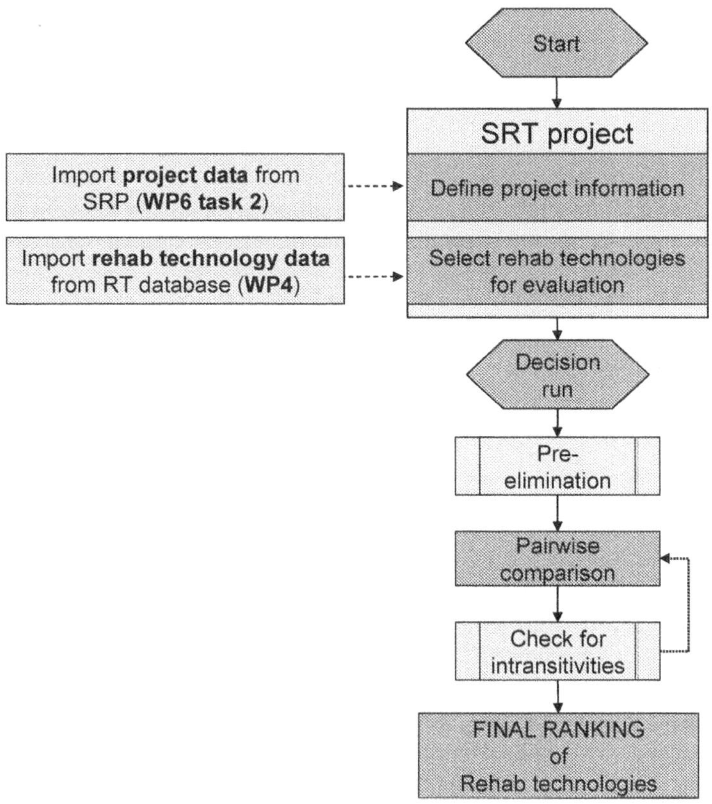

Figure 6.21: Workflow structure chart of CARE-S – SRT

Software design

Beside typical data management facilities, the general design of the CARE-S – SRT software consists of two main screens. The first one provides information on the rehabilitation project, including pipe characteristics, failure specification, and description of the environment (Figure 6.22). Here, the pre-selection of suitable technologies is carried out.

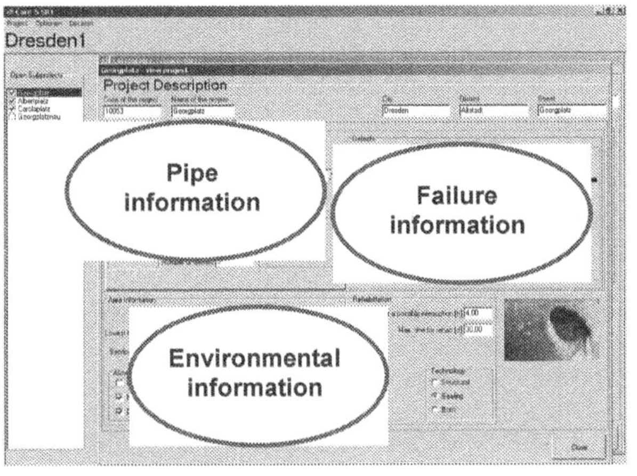

Figure 6.22: Project description

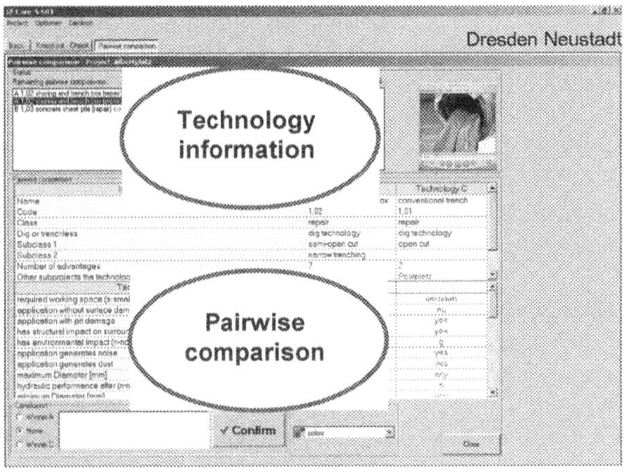

Figure 6.23: Ranking

On the second screen, the remaining rehabilitation technologies are listed with their advantages and disadvantages, and the ranking process is carried out (Figure 6.23). A more detailed description of the software is included in the software handbook.

References

ATV – Abwassertechnische Vereinigung
M 143-2: Inspektion, Instandsetzung, Sanierung und Erneuerung von Abwasserkanälen und -leitungen. Teil 2: Optische Inspektion (1999)
M 149: Zustandserfassung, -klassifizierung und –bewertung von Entwässerungssystemen außerhalb von Gebäuden. (1999)

Baur, R. (2003): Bestimmung kosteneffizienter Rehabilitationsprojekte. Dargestellt am Beispiel städtischer Wasserrohrnetze. Dresdner Beiträge zum Stadtbauwesen. TU Dresden. 152p + Annex

Baur, R., W. Zielichowski-Haber and I. Kropp (2004): Statistical Analysis of Inspection Data for the Asset Management in Sewer Networks. 19th EJSW on Process data and integrated urban water modelling. Lyon, France. March 2004

Baur, R., I. Kropp und R. Herz, Eds. (2003): Rehabilitation management of urban infrastructure networks. Proceedings of the 17th European Junior Scientist Workshop - EJSW - (3.-7. Sept. 2003), Dresden, 211 pp ISBN 3-86005-403-1

Baur, R., P. Le Gauffre and S. Sægrov (2003): Multi-Criteria Decision support for drinking water network rehabilitation. JWST: Water Supply. 3, 1-2, 43-50.

Baur, R., R. Herz, and I. Kropp (2003): Choosing the right sewer rehabilitation technology. D16 report. CARE-S (Computer Aided Rehabilitation of Sewer and storm water networks), EU project under the 5th Framework Programme, contract n°EVK1-CT-2002-00106, 34pp + Annex

Baur, R., R. Herz, and I. Kropp (2005): Multi-Criteria Decision support methodologies - WP 6 conclusive report. CARE-S (Computer Aided Rehabilitation of Sewer and storm water networks), EU project under the 5th Framework Programme, contract n°EVK1-CT-2002-00106

Baur, R. and R. Herz (2002): Selective inspection planning with ageing forecast for sewer types. Journal of Water Science and Technology. 46, 4-7, 389-396.

Baur, R., and I. Kropp (2002): Selecting projects for efficient water network rehabilitation. DMinUCE, London, UK

BMB+F (2002): Entwicklung eines allgemein anwendbaren Verfahrens zur selektiven Erstinspektion von Abwasserkanälen und Anschlussleitungen. Teil A: wissenschaftliche Untersuchungen. Institut für Siedlungswasserwirtschaft. RWTH Aachen

Diab Y., Morand D. (2000). Muticriteria choice of rehabilitation techniques for small urban sewers VI International pipeline construction congress (proceedings). pp 588-594. 2000

DIN EN 752-5 (1997): Drain and sewer systems outside buildings. Part 5: Rehabilitation

EU (2000): http://europa.eu.int/comm/environment/water/water-framework/

Gero, J.S., A.D. Radford und N.S. Murthy (1982): What If? Exploring the consequences of design and performance decisions in computer-aided design. CAD82, Butterworths, Guildford. UK. p. 633-645.

Grunwald, G. (1997): Wirtschaftlichkeitsuntersuchungen bei Kanalsanierungen. 158 p + Annex. Gelsenkirchen, Germany

Herz, R. (2002): Software for strategic network rehabilitation and investment planning. International conference. Computer Aided Rehabilitation of Water Networks CARE-W. Dresden, pp.65-84

Herz, R. (1998): Exploring rehabilitation needs and strategies for drinking water distribution networks. J Water SRT – Aqua 1998; 74: 275 - 283

Herz. R. (1996a): Ageing processes and rehabilitation needs of drinking water distribution networks. J Water SRT - Aqua 45. 5. pp.221-231.

Herz. R. (1996b): Dégradation et renouvellement des infrastructures : un modèle de survie par cohorte. Flux, 23, 21-36

Herz. R. (1995): Alterung und Erneuerung von Infrastrukturbeständen – ein Kohortenüberlebensmodell. Jahrbuch für Regionalwissenschaft, 14/15, 5-29

Herz, R. and I. Kropp (2002): WP4 – Strategic planning and investment: Development of the "Rehab strategy manager" software. D10 report. CARE-W (Computer Aided Rehabilitation of Water networks), EU project under the 5th Framework Program, contract n°EVK1-CT-2000-00053, 26 p. + Annex

Herz, R., R. Baur, A. Lipkow and I. Kropp (2003): WP4 – Strategic planning and investment: Development of the "Rehab Strategy Evaluator" software. D11 report. CARE-W (Computer Aided Rehabilitation of Water networks), EU project under the 5th Framework Program, contract n°EVK1-CT-2000-00053, 18p + Annex

Herz. R. and K. Hochstrate (1987): Erneuerungsstrategien für städtische Infrastrukturnetze. Jahrbuch für Regionalwissenschaft, 8, 67-105.

Herz. R. and K. Krug (2000): Sanierungsbedarf und Sanierungsstrategien für Abwasserkanäle. Beitrag zum 11.Leipziger Bau-Seminar.

Hochstrate, K. (1986): Interaktives lösungsraumorientiertes Entscheidungsverfahren für Infrastrukturinvestitionen. Schriftenreihe des Instituts für Städtebau und Landesplanung H.19, Universität Karlsruhe, Germany. 157 p.

Hochstrate, K. (1999): Substanzwertorientierte Zustandsklassifizierung von Kanälen – das Bietigheimer Modell. KA – Wasserwirtschaft, Abwasser, Abfall, 46, 2, 213-217

Hochstrate, K. (2000): Sewer status assessment by rehabilitation priority, intrinsic value and functionality as basis for forecast supported inspections and predictive rehabilitation planning. 6. Internationaler Kongress Leitungsbau. Hamburg. 548-557

KANEW: www.tu-dresden.de/biwiss/stadtbau/KANEW.html

Kleiner, Y. (2001): Scheduling inspection and renewal of large infrastructure assets. Journal of infrastructure systems. 7, 4, 136-143

Kropp, I. and R. Baur (2002): Computer aided rehab budget allocation - a case study from Germany. IWA ManProWU Conference, Praha, Czech Republic, pp 132-138.

Le Gat, Y. (2003): Degradation Models for Drain and Sewer Pipelines Parameterised with CCTV Inspection Data. CARE-S. Internal paper. unpublished.

Le Gauffre P., R. Baur and K. Laffréchine (2002b): Survey of multi-criteria techniques and selection of relevant procedures. D7 report. CARE-W (Computer Aided Rehabilitation of Water networks), EU project under the 5th Framework Program, contract n°EVK1-CT-2000-00053, 26 p. + Annex

Le Gauffre P., R. Baur, K. Laffréchine and M. Schiatti (2002a): Survey of multi-criteria techniques and selection of relevant procedures. D7 report. CARE-W (Computer Aided Rehabilitation of Water networks), EU project under the 5th Framework Program, contract n°EVK1-CT-2000-00053, 26 pp + Annex

Le Gauffre, P., C. Joannis, C. Gibello and D. Breysse (2002c): Performance indicators and decision support for rehabilitation of sewer networks. Contribution of the French R&D RERAU program. International conference on Sewer operation and maintenance. Bradford

Le Gauffre, P., C. Joannis, D. Breysse, C. Gibello and J.J. Desmulliez (2004): Gestion patrimoniale des réseaux d'assainissement urbains. Guide méthodologique. Paris: Lavoisier Tec&Doc, 416p, ISBN 2-7430-0748-6.

Ministerium für Umwelt und Verkehr Baden-Württemberg (=MUV-BW 2000): Leitfaden für kostenminimierende Instandhaltung von Kanalnetzen. Stuttgart. 49 p.

Plenker, T. (2002): Computer aided decision support on choosing the right technology for sewer rehabilitation. Water Science and Technology, 46, 6-7, 403-410

Plenker, T. (2003): Multikriterielles Auswahlverfahren zur Bestimmung der bestgeeigneten Sanierungstechnik für individuelle Abwasserkanäle. PhD thesis, TU Dresden, 187p.

Roy B., and D. Bouyssou (1993): Aide multicritère à la décision: méthodes et cas, Economica, Paris.

Roy, B. (1985): Méthodologie multicritère d'aide à la décision. Economica. Paris

Schmidt, T. (2002): Rechnergestütztes Abwägungsverfahren für Infrastruktur-Rehabilitationsprojekte mit einer Anwendung auf die Auswahl des besten Sanierungsverfahrens für einen Abwasserkanal unter spezifischen örtlichen Randbedingungen. Diploma thesis, TU Dresden, 98p.

Schulz, N., R. Baur and P. Krebs (2004): Integrated modelling for the evaluation of infiltration effects. 10th UDM conference, Dresden, Germany

Sjøvold, F., I. Selseth, and S. Sægrov (2005): Selection of pipes for CCTV inspections to calibrate LTP model. CARE-S memo. 9p.
Stein, D. (1998): Instandhaltung von Kanalisationen. Ernst & Sohn. Berlin. 941p.
Strassert, G. (1984): Entscheiden über Alternativen ohne Super-Zielfunktion: Schrittweise und interaktiv. Eine Neuorientierung multidimensionaler Entscheidungstechnik. IfR Diskussionspapier Nr.14, Institut für Regionalwissenschaft, Universität Karlsruhe
Strassert, G. (1995): Das Abwägungsproblem bei multikriteriellen Entscheidungen. Grundlagen und Lösungsansatz unter besonderer Berücksichtigung der Regionalplanung. Peter Lang Verlag, Frankfurt am Main.
Strassert, G. (2000): The balancing principle, strict superiority relations, and a transitive overall final order of options. Diskussionspapier Nr.34, Institut für Regionalwissenschaft der Universität Karlsruhe. 19 p.
Vincke, Ph. (1992): Multicriteria decision aid. John Wiley. Chichester, UK, 154p
WRc (2001): Sewerage rehabilitation manual. 4th edition (SRM4)

7

Development of the wastewater network rehabilitation manager

Roger Hurley, Project Manager, Nick Orman, Senior Consultant, Joanne Hulance, Consultant Engineer, Kevin Taylor, Consultant Engineer, Jeanette Cooke, Business Scientist

With contributions from: Paul Brettell, Severn Trent Wastewater, John Hateley, Severn Trent Wastewater, Tony Lowles, United Utilities

7.1 BACKGROUND

The overall aim of the CARE-S project is to develop methods and software that will enable engineers to define and implement effective management of their sewer networks, *"rehabilitating the right pipes at the right time"*. CARE-S provides a methodology for rehabilitation planning consistent with the procedure described in European Standard EN752-5:1997. It consists of a step-by-step procedure, a number of software tools and a *Wastewater Network Rehabilitation Manager* that guides the user through the procedure and provides the means of integrating the various software tools.

The Rehabilitation Manager is more than just a software application linking a collection of analysis tools. It is a versatile and powerful application, facilitating the efficient use of the CARE-S toolkit. At its simplest level it provides a central storage area for appropriate asset data and analysis results. The Rehabilitation Manager also facilitates data manipulation, allowing the user to view the data both geospatially and in tabular form. It manages the data interactions between all the CARE-S tools and produces appropriate reports to facilitate the development of a rehabilitation plan. Finally, the Rehabilitation Manager provides interactive guidance for the selection of the appropriate analysis method, in accordance with the CARE-S procedure.

Work Packages WP1 to WP6 of the CARE-S project have focused on the identification, design and development of the tools whose objectives are to model and predict typical problems, such as structural and hydraulic failures, identify possible rehabilitation techniques, quantify the direct and indirect costs of the failures and rehabilitation options, and identify appropriate rehabilitation strategies. WP7 is concerned with producing the software to enable the CARE-S tools to be applied methodically to the rehabilitation planning process.

© 2006 The Authors. *CARE-S* edited by Sveinung Saegrov. ISBN: 9781843391159. Published by IWA Publishing, London UK

The specific objectives of WP7 are as follows:

- To specify the data input/output and storage requirements for the integrated CARE-S package;
- To produce a specification for software that will optimally integrate the use of the tools defined in CARE-S;
- To deliver working prototype software that will manage the information needed by, and results generated from, the CARE-S tools, enabling the integrated CARE-S package to identify optimum rehabilitation strategies.

To develop the CARE-S procedure, the requirements of the planning procedure described in EN752-5 must be clearly understood and the individual CARE-S tools must be thoroughly evaluated in terms of the benefits they bring to the process, their data requirements and the inter-relationships between them. The role of each tool must be determined and a procedure for optimum integration developed. The procedure must clarify how to utilise the CARE-S toolbox most efficiently in order to address a given problem or circumstance.

The key performance indicators and data flows must be developed and specified as a prerequisite to development of the software. The structure of the Rehabilitation Manager software will depend upon the data inputs, outputs and reports that are required and/or produced by the CARE-S tools. It is therefore essential that this information be thoroughly specified.

The Rehabilitation Manager software must support the integration of the tools and the data flows between them. The software will allow automatic and manual data input and will store information (results of analysis) that can, for example, enable electronic write-back to corporate data storage systems. A manual will be produced to support the use of the software.

7.2 SCIENTIFIC METHODOLOGY APPLIED

The starting point for this work package was to understand the overall requirements of the rehabilitation planning procedure and to assess the position within this procedure of the various tools that comprise the CARE-S tool-kit. The nature and benefits of each tool needed to be fully understood in order to construct a protocol for the integrated application of the tools. Once this protocol for operation had been defined, the task of integrating the software and modelling tools could begin. Data input, output, storage and handling were specified.

The next stage was to design and specify the Rehabilitation Manager software such that the necessary data flows could be accommodated. It was important to structure and configure this software so that it would provide an effective conduit between corporate data systems and the CARE-S suite of tools. Finally, the software was created, tested and delivered as a fully functional software application.

The more detailed methodology adopted to deliver the CARE-S procedure and Rehabilitation Manager software is outlined below.

7.2.1 The CARE-S Procedure

A general framework for rehabilitation planning is provided in EN752-5; this is summarised in Figure 7.1. In developing CARE-S, it was necessary to ensure that the framework was flexible enough to address a range of European needs, whilst still allowing the user to focus on local or national requirements. Major issues included the variability in data quality and availability, and differences in planning objectives. CARE-S must be viable across Europe and not just for those sewerage service providers who hold good-quality data. Furthermore, the tools must offer benefits for small municipalities as well as the larger, private companies.

Having established the general framework, the next step was to map the CARE-S tools onto it and establish the conceptual links between them. This is shown in Figure 7.2.

It was then possible to identify the step-by-step process for applying CARE-S and to specify the data flows into and out of CARE-S and, internally, between the CARE-S tools. All these data flows are facilitated by the Rehabilitation Manager.

Details of the development of the CARE-S Procedure are presented in CARE-S report D20, "The CARE-S Procedure".

The Rehabilitation Manager thus facilitates the use of the various CARE-S tools within the general procedure. One of the key features is that decision-making information can be passed from one tool to the next. This is critical to successful tool integration and interaction. The development of the software is explained in the next section.

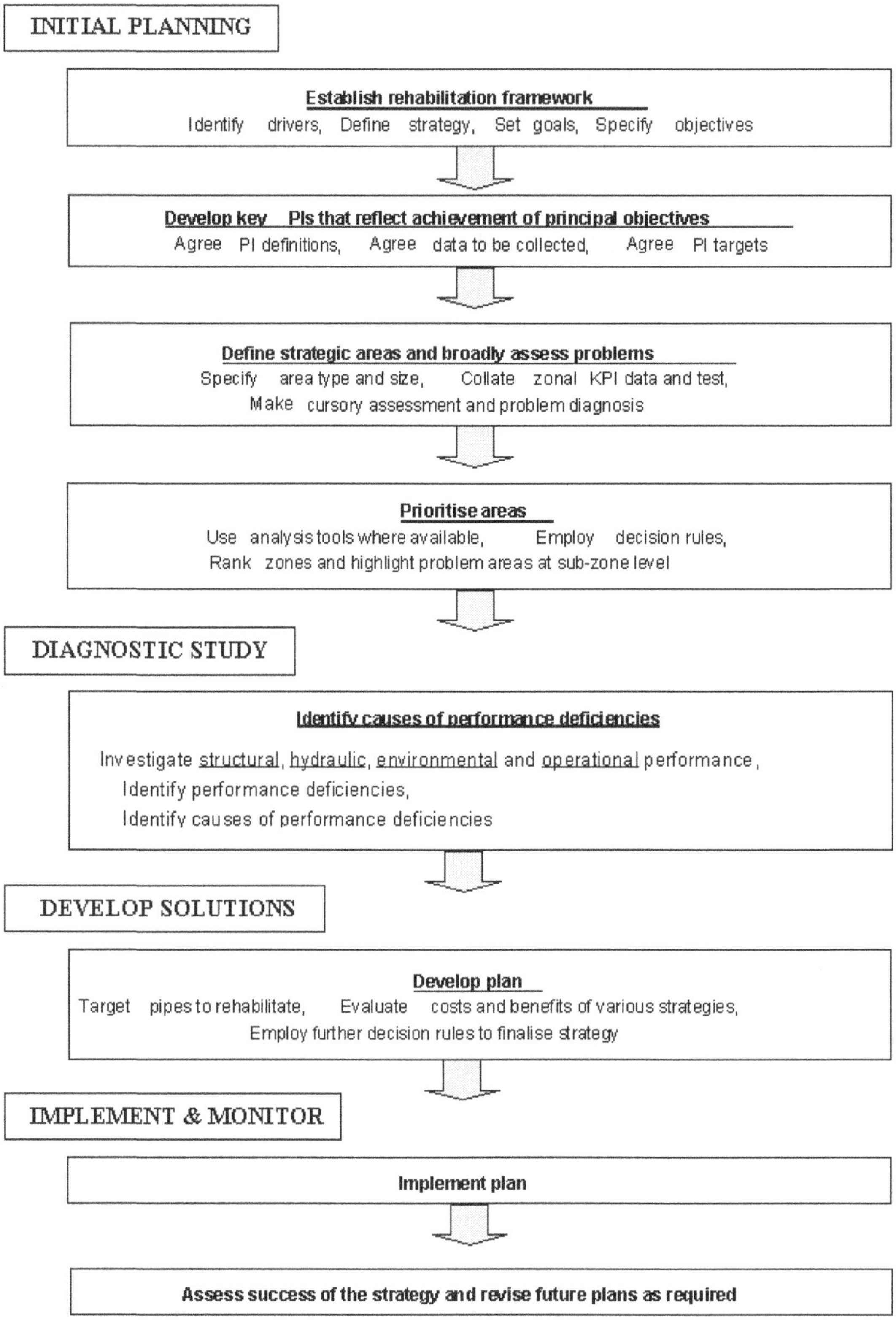

Figure 7.1 Generic Rehabilitation Planning Procedure

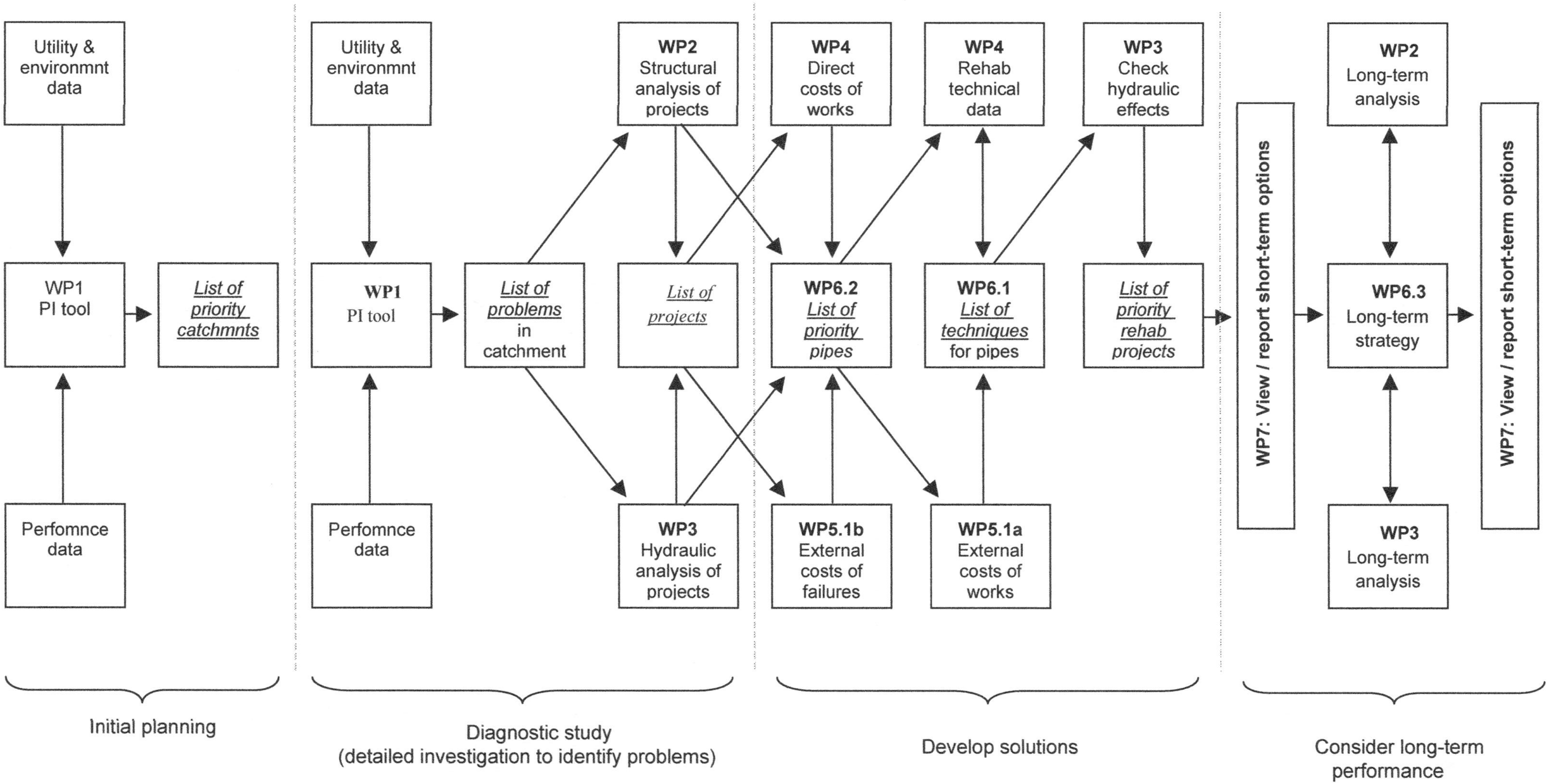

Figure 7.2 Conceptual links between work packages

7.2.2 The Wastewater Rehabilitation Manager Software

The methodology for developing the Rehabilitation Manager software was defined by the following requirements:

- It must provide a central depository for the data required by the CARE-S tools used in the rehabilitation planning process;
- It must manage the flows of data into and out of CARE-S, and between the CARE-S tools, presenting data in a format acceptable to the tool;
- It must provide users with outputs of results where these are not provided by the tools;
- It must guide users through the rehabilitation planning process;

The software development therefore involved the following steps:

- Specification of data input/output and storage requirements for the integrated CARE-S package;
- Production of a specification for the software that would integrate the use of the tools defined in CARE-S in an optimal manner;
- Delivery of working software which would manage the information needed and results generated from the CARE-S tools, enabling the integrated CARE-S package to identify optimum rehabilitation strategies.

The Rehabilitation Manager would thus lie at the centre of a versatile and powerful software application, incorporating a variety of rehabilitation planning tools.

Details of the development of the Rehabilitation Manager software are presented in CARE-S report D21, "User interface for the CARE-S wastewater rehabilitation manager software".

7.3 SCIENTIFIC RESULTS OBTAINED

There were two specific outputs from WP7:

- The CARE-S Procedure for Rehabilitation Planning; and
- The Wastewater Network Rehabilitation Manager Software.

These two outputs are described in more detail below.

7.3.1 The CARE-S Procedure

Introduction

Efficient planning requires the rehabilitation engineer or planner to be in possession of a substantial amount of background knowledge and experience on the types of problems faced, current performance and possible effective solutions. The engineer must be aware of the objectives of rehabilitation for each problem he faces, and apply sound judgement using all the tools at his disposal in an appropriate manner. This places a huge burden on the engineer when many rehabilitation methods are feasible and there are many solutions to improve service delivery to customers. The engineer can narrow his search for an acceptable solution by using strategic level data and knowledge of historical performance. However, scheduling rehabilitation over, say, a 20-year planning horizon is a complex multivariate problem and the engineer must be guided by a suite of useful analysis tools to help him choose the most cost-effective rehabilitation options from those available.

A general framework for developing this rehabilitation plan is provided by EN752-5 (as described in section 0 above) and CARE-S provides a software application which contains a set of 'tools' to apply within the framework. The software guides users through the EN752-5 procedure and enables them to apply the tools in a logical and integrated manner, passing data from one tool to the next and managing the overall process. The stages of the procedure and the CARE-S tools utilised at each stage are described below (see also Figure 7.1 and Figure 7.2 above).

Mapping the CARE-S tools onto the EN752-5 framework

Initial planning

At the initial planning stage the objective is to identify problem catchments (or sub-catchments). This is achieved by analysing overall system performance (i.e. at company or regional level) using the WP1 (performance indicator, or PI) tool. It requires inputs of the basic utility and environment data plus performance data. The result is a general understanding of the nature and severity of the problems, so that planners can generate a list of priority catchments and/or subcatchments for attention.

The PI tool provides definitions of appropriate performance indicators together with algorithms for calculating them. Performance data are provided by users and entered via the Rehabilitation Manager's user interface for storage in the CARE-S database. The rehabilitation manager supplies the PI tool with the data it requires and the tool, in turn, returns PI values to the rehabilitation manager for storage or reporting.

Diagnostic study

The first catchment or subcatchment is then selected for investigation. The WP1 PI tool is used again, but this time at a more detailed level than before, to identify problems within the area under investigation. The result is a list of performance problems in the area.

The WP2 (structural performance) and WP3 (hydraulic performance) tools are then used to analyse the structural/operational and hydraulic/environmental performance of the sewer system, to understand the causes of the problems identified. This produces a list of structural and hydraulic deficiencies to be solved by rehabilitation projects.

Develop solutions

The WP4, 5 and 6 tools are used at this stage in the procedure. WP4 provides a catalogue of rehabilitation techniques, enabling techniques to be selected and their direct costs estimated. WP5 provides socio-economic cost data (the indirect costs of sewer failures and rehabilitation works). WP6 provides decision-making support at various stages in the rehabilitation planning process, enabling users to identify the most urgent problem pipes and select the most appropriate rehabilitation options for these pipes.

The pipes responsible for the structural and hydraulic deficiencies are analysed by the WP6.2 (pipe selection) tool to produce a list of priority pipes for the area. WP6.2 requires inputs from WP2 (structural performance), WP3 (hydraulic performance), WP4 (direct costs of possible rehabilitation techniques) and WP5.1 (indirect costs of failures and rehabilitation works). The Rehabilitation Manager passes the list of pipes produced by the WP6.2 tool to WP4 and WP5.1 where pre-elimination of rehabilitation techniques is carried out and the direct and indirect costs of the solutions are assessed. The results are returned to the WP6.1 (technology selection) tool and, using this information, rehabilitation techniques for the priority pipes are identified.

The user must select a rehabilitation technique for each pipe from the options presented by the WP6.1 tool and must check the hydraulic effects of the proposed works, using the WP3 tools and proprietary hydraulic modelling software applications.

It should be noted that hydraulic modelling with proprietary applications is undertaken outside CARE-S. The Rehabilitation Manager provides links to the hydraulic model's input and output data files, which operate as external satellites of the CARE-S database. The Rehabilitation Manager converts the data from MOUSE, SWMM and Infoworks formats to 'CARE-S' format for use by the tools, effectively integrating these hydraulic modelling packages into CARE-S.

The list of priority projects thus produced will support the rehabilitation engineer in developing the short-term rehabilitation plan, enabling the most urgent works to be set in progress at the earliest opportunity. The next stage in the procedure is to develop a long-term plan.

Consider long-term performance

Further evaluation of the proposed solutions, to develop a long-term (20 year) strategy, is then undertaken by the WP6.3 (rehabilitation strategies) tool in conjunction with the WP2 (structural) and WP3 (hydraulic) tools. The Rehabilitation Manager provides the required data to the WP6.3 tool and stores any results required for composite reporting, in the database.

Using CARE-S

Having mapped the CARE-S tools onto the EN752-5 framework, a detailed step-by-step guide to the use of the software was produced. This links each step in the EN752-5 procedure with a CARE-S task, or number of tasks in the manner shown in Table 7.1:

Table 7.1 Using CARE-S

EN752-5 task	***CARE-S Tasks***
STAGE 1 – INITIAL PLANNING	Stage 1 is undertaken at a high level (e.g. the entire WSP area) to set the ground-rules for the rehabilitation studies and establish priorities between individual catchments and sub-catchments for the studies.
Establish the framework	
Identify reasons why rehabilitation is being considered (flooding, pollution, blockages, etc.) and rank them in order of importance.	This is obtained from the users' general knowledge of his sewer systems
Identify PIs which relate to these identified reasons for rehabilitation.	CARE-S will have no direct input here, but the WP1 list of PIs is a useful reference of possible PIs.
Agree planning horizon of rehab plan	
Define PIs	
Set triggers & targets for PIs (structural, hydraulic, environmental & operational, as relevant)	
Establish robust data sources	
Define the extent of the area to be assessed and make a broad assessment of the problems	
Catchment or sub-catchment studies? Users must choose between a data-intensive detailed study of a small area or a more general study of a large area.	The decision between investigations at catchment or sub-catchment level can be done by the WP1 PI tool, but in reality is more likely to be done outside CARE-S, based on user's knowledge.

	Establish a CARE-S 'project' for the entire area and divide into catchments or sub-catchments for comparison of high level PIs. *Import asset data from external sources* to WP7 database (these may be GIS, or CSV files, or files already created for any of the CARE-S tools). At this stage, a single 'project' will be created for the entire company area. The catchments (or sub-catchments) will form datasets within the projects. When carrying out the detailed rehabilitation studies (that is, EN752-5 'detailed investigations') on the catchments, a separate 'project' will be created for each catchment.
	Import trigger and target values for each PI from external sources to WP1 PI tool. Many PI's are area-specific (not pipe-specific) so the areas must be defined before the PI's are calculated.
Assess current performance in order to define precise scope of study and set overall programme including other studies.	*Export asset and performance data from WP7 database* to WP1 PI tool. The WP7 Rehabilitation Manager can produce an asset summary report at this stage.
	Run WP1 PI tool
	Import PI values for each catchment or sub-catchment from WP1 PI tool to WP7 database. The PI tool will save results directly to the WP7 database.
Decide on level of detail of study (simple, structural only, hydraulic/environmental only, etc).	WP7 Rehabilitation Manager generates reports on the performance of each catchment or sub-catchment. Users must review the PI data and decide on the level of study required in each sub-catchment. This decision is taken outside of CARE-S.
Prioritise areas	
Assess the PI data for all catchments and sub-catchments under consideration and set priorities & programmes for each study.	Users must review the PI data and decide on priorities for rehabilitation studies. This decision is taken outside of CARE-S.
	…… etc.

7.3.2 The Wastewater Rehabilitation Manager Software

The Rehabilitation Manager lies at the core of CARE-S. It has three key attributes:

- The ability to store key data in a central repository;
- The ability to use a number of rehabilitation planning tools in an integrated manner;
- The ability for the rehabilitation planning tools to interact and share data.

The software consists of an MS "Access 2000" database application, written in Visual Basic 6.0, with the following functionality:

- It provides a central storage area and reference point for essential CARE-S data;
- It accepts *user-validated* input data in a pre-defined format. Data source fields are mapped to fields within the CARE-S database according to the specification of the declared input data source;
- It provides automatic conversion of data item units to standard CARE-S convention;
- It performs rudimentary data validation including duplication, data type and range checking;
- It allows user interaction with a pre-determined range of tools under the CARE-S umbrella;
- It interacts with a pre-determined range of tools at the user's request by:
  a) Creating the necessary input file(s) from the data stored in the CARE-S database to enable the tool to be run; and
  b) Accepting and storing necessary output data on completion of tool use.

The general architecture of the CARE-S Rehabilitation Manager software is shown in Figure 7.3.

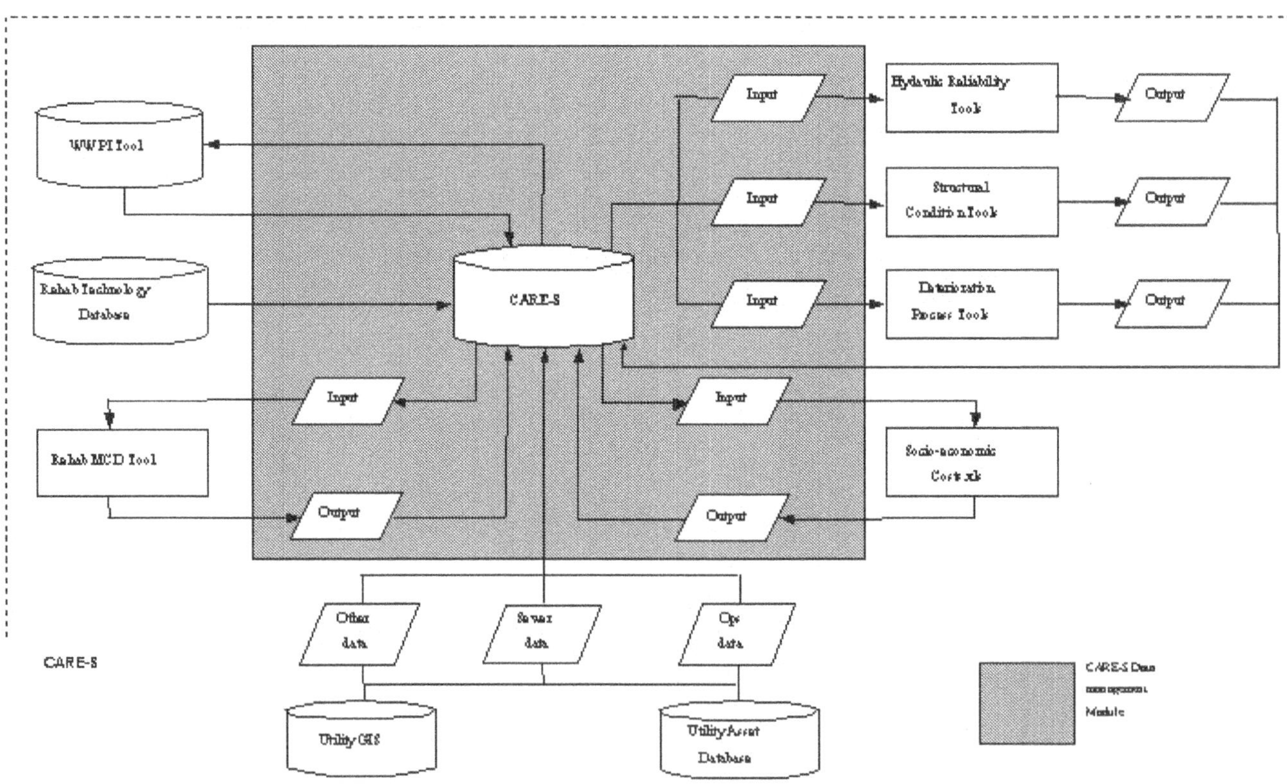

Figure 7.3 Rehabilitation Manager architectural design

The CARE-S central database

The central database consists of a series of tables which are used to store information from a number of different input sources. Each table contains specific types of information, such as pipe based data, zonal, geospatial or hydraulic information. Other tables hold information specific to a *project* or a *dataset*, including analysis results. The necessary links between input and results data are maintained to allow proper reporting of analysis history and results to the user.

Each database table contains fields which relate to the various parameters or data items used by each tool. Many of these fields, such as pipe diameter, length and material, are used by a number of different tools. Others are unique to a specific tool. It has been necessary to adopt a standardised set of parameter properties in order that CARE-S can communicate with all analysis tools.

At the request of the user, the Rehabilitation Manager will create the necessary input file(s) for an analysis tool according to a pre-defined specification held internally. These files may then be used to run the analysis tool within (or outside) the CARE-S framework. Similarly, the Rehabilitation Manager will enable the results of the analysis to be exported back to CARE-S and stored as necessary on completion of the analysis task.

User interface

Menu structure

Figure 7.4 shows the menu options and toolbar of the CARE-S Rehabilitation Manager software. The title bar details the name of the project that is currently open (in this example, 'Demo'), the active dataset name (i.e. 'V984-01') and the active dataset description (in this case, 'No 1').

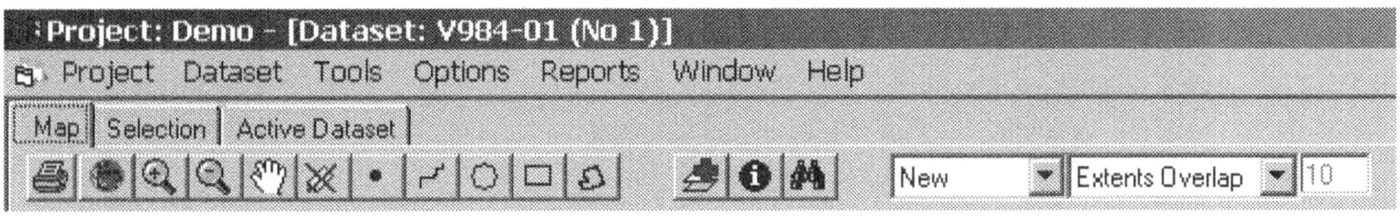

Figure 7.4 CARE-S menu options and toolbar

The menu bar provides access to the range of functionality available in CARE-S.

- The **Project** menu allows the user to open, close and save projects, with or without new names, and give access to the Project Properties dialog, a summary of the key features of the current (open) project.
- The **Dataset** menu allows the dataset properties to be modified, including the selection of assets which form the active dataset. This requires the user to make a selection of assets first by using the GIS selection features or "find" facility.
- The **Tools** menu gives access to all the functions required to interact with the various CARE-S tools, including importing and exporting data, modifying input files, and invoking the tool application.
- The **Options** menu allows the user to define a number of personal preferences on display of CARE-S project information, as well as providing access to a number of miscellaneous functions.
- The **Reports** menu gives access to a range of tabular and geo-spatial report facilities allowing representation of results from all CARE-S tools.
- The **Window** and **Help** menus complete the menu bar, providing a number of standard functions similar to those in other Windows applications, including access to CARE-S help facilities and registration/licence details for the software installation.

The GIS viewer

The GIS viewer is the principal feature of the CARE-S Rehabilitation Manager's user interface. It allows:

- The geographical display and interpretation of asset data
- The selection and manipulation of geographic elements in datasets
- The representation of thematic maps including reporting of CARE-S analysis results

An example of the use of the viewer is shown in Figure 7.5.

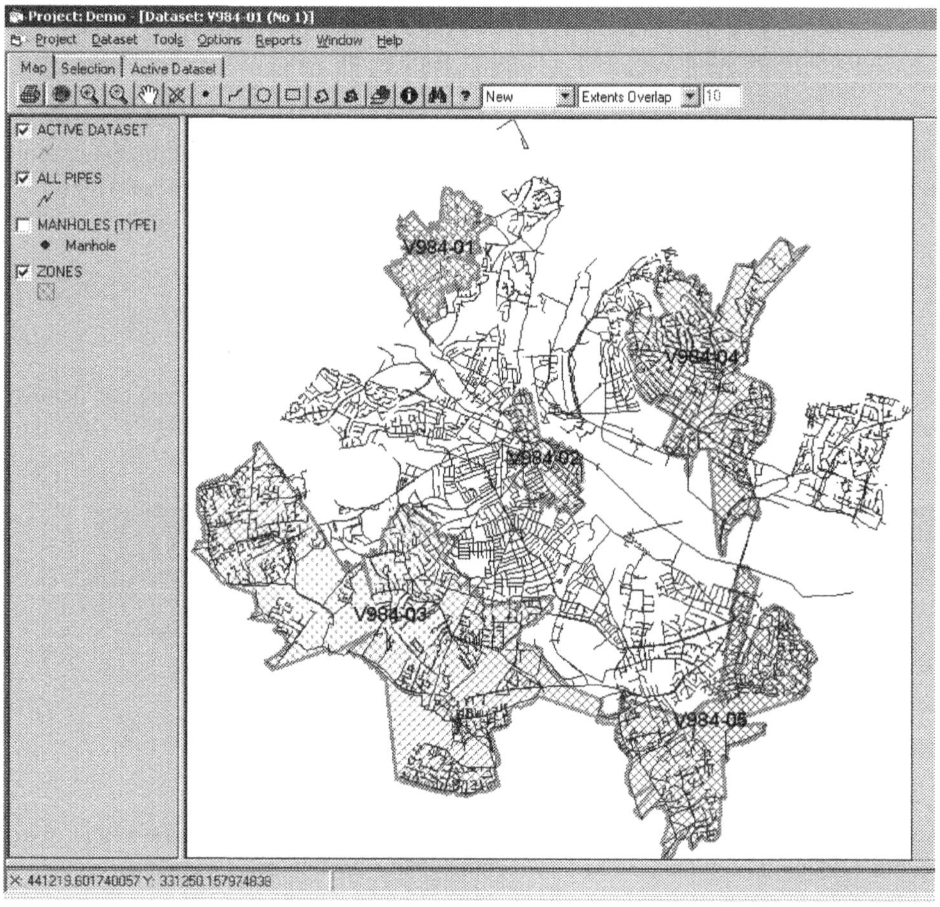

Figure 7.5 The CARE-S GIS Viewer

As with other standard GIS applications, information is structured in layers, each of which may be viewed or 'switched off' by means of the check boxes in the left-hand pane of the GIS window. When data are imported, a standard colour scheme, line style and/or object are used to display different assets. Note that in Figure 7.5 the manhole positional data (represented by small filled blue circles) has been switched off for reasons of clarity. In CARE-S, the active dataset (that which is of interest to the user at the current time) is highlighted in a different colour (in this case orange). Datasets are denoted by polygons filled by hatching of the same colour. The boundaries of these polygons are constructed automatically. The name of the dataset is printed at the centroid of the polygon. It can be seen in Figure 7.5 that the asset data imported consists of a number of non-contiguous datasets, all of which may be analysed individually by CARE-S tools.

Tabular views

In addition to the spatial representation of data and results, there are tabular views for users of datasets without geospatial referencing. Although the functionality is necessarily more limited than with the GIS viewer, these views contain the essential asset information and provide the means for manual inspection before proceeding with making a selection of assets for further analysis or reporting. An example of one such view is given in Figure 7.6. Each record denotes a sewer, with details such as upstream and downstream node names, length of section, construction material, and so on.

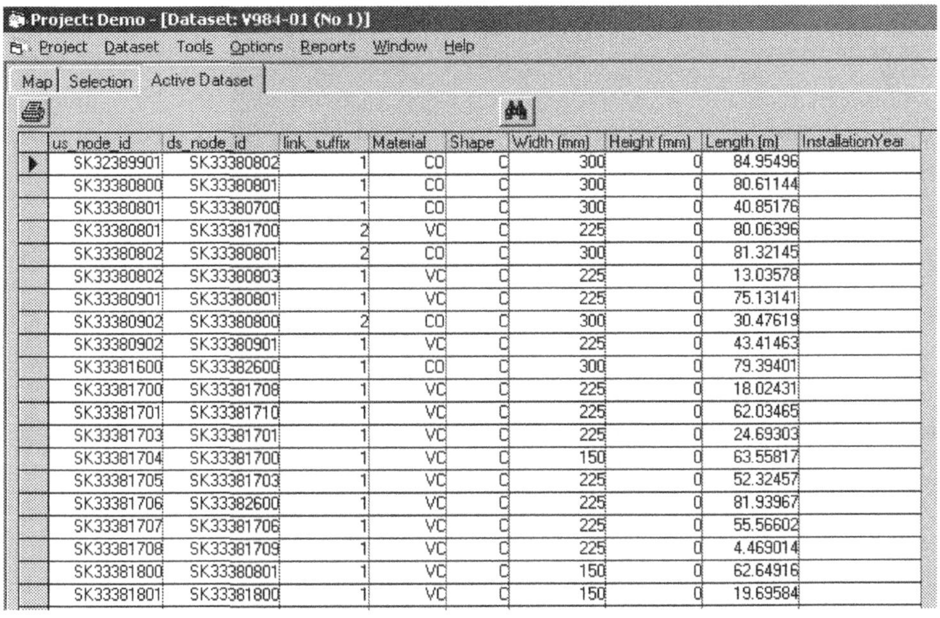

us_node_id	ds_node_id	link_suffix	Material	Shape	Width (mm)	Height (mm)	Length (m)	InstallationYear
SK32389901	SK33380802	1	CO	C	300	0	84.95496	
SK33380800	SK33380801	1	CO	C	300	0	80.61144	
SK33380801	SK33380700	1	CO	C	300	0	40.85176	
SK33380801	SK33381700	2	VC	C	225	0	80.06396	
SK33380802	SK33380801	2	CO	C	300	0	81.32145	
SK33380802	SK33380803	1	VC	C	225	0	13.03578	
SK33380901	SK33380801	1	VC	C	225	0	75.13141	
SK33380902	SK33380800	2	CO	C	300	0	30.47619	
SK33380902	SK33380901	1	VC	C	225	0	43.41463	
SK33381600	SK33382600	1	CO	C	300	0	79.39401	
SK33381700	SK33381708	1	VC	C	225	0	18.02431	
SK33381701	SK33381710	1	VC	C	225	0	62.03465	
SK33381703	SK33381701	1	VC	C	225	0	24.69303	
SK33381704	SK33381700	1	VC	C	150	0	63.55817	
SK33381705	SK33381703	1	VC	C	225	0	52.32457	
SK33381706	SK33382600	1	VC	C	225	0	81.93967	
SK33381707	SK33381706	1	VC	C	225	0	55.56602	
SK33381708	SK33381709	1	VC	C	225	0	4.469014	
SK33381800	SK33380801	1	VC	C	150	0	62.64916	
SK33381801	SK33381800	1	VC	C	150	0	19.69584	

Figure 7.6 Tabular view example

Data flows and tool interaction

The CARE-S database is a cornerstone in the integration of the set of tools which help users to develop their rehabilitation plans and strategies for the years ahead. This is made possible by the facility to move data between the database and the tools, and vice-versa. The main benefit provided to CARE-S by the Rehabilitation Manager is the power and flexibility of data handling and analysis.

Importing data

Data may be imported into CARE-S from any of three sources, namely:

- CSV data files, e.g. asset database or maintenance records;
- GIS data files, e.g. ArcView 'shape' files;
- External analysis tools, e.g. 'Infoworks', 'MOUSE' or 'SWMM' hydraulic model files.
- The process of importing data into a CARE-S project is summarised in Figure 7.7.

The first decision the user must make is the nature of the import. Are the data from an asset inventory or from a hydraulic model? If it is the former, it is assumed that the data are unverified and, provided that a specification for data import has been defined, these data will be imported to the CARE-S project database as an unverified dataset. In the case of hydraulic model data, the user must specify whether the data to be imported are fundamental model data or the results of a flow simulation, and confirm any relationship with existing model data already held in that CARE-S project from an earlier import.

It should be noted that asset data and hydraulic model data are the only types of data that the user will be expected to import using the 'Import' menu. Other data, e.g. PI values, structural performance, socio-economic factors, etc., will be imported from the relevant tools by the Rehabilitation Manager on user request using the 'Tools' menu.

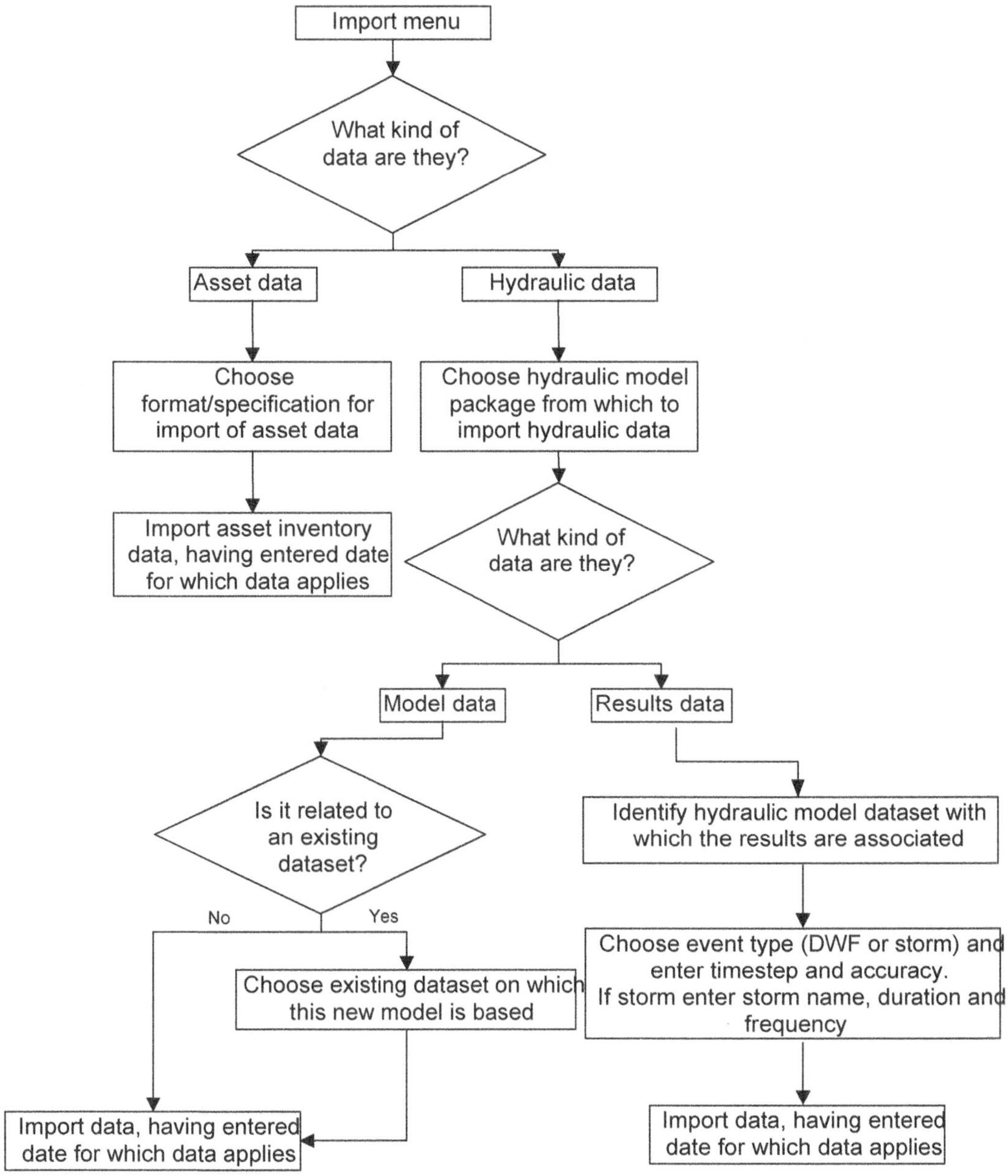

Figure 7.7 Process chart for data import

Data validation

Any analysis undertaken using the CARE-S software is only ever as good as the quality of the data used in that analysis. It remains the user's responsibility to ensure that valid data are available for import, that is, the data are arranged in a consistent and definable manner, data types are correct and data values are meaningful.

That said, CARE-S does provide some rudimentary data validation. For instance, the units (m, mm, MPa, etc.) and data types (text, integer, binary [yes/no], etc.) are defined for the fields in the central database tables. On import to the database tables, the data are automatically converted to the standard units required by the tables. Conversely, on export, the data are converted to the units required by the individual CARE-S tools (although it is noted that, in most cases, the database units are consistent with those required by the tools). The software also checks the data type as the import is carried out and checks that the type matches that of the recipient field in the CARE-S database. The user is notified of any data format violations during this validation.

Running the CARE-S tools

Most interactions between users and the tools included in CARE-S are managed via the 'Tools' menu shown in Figure 7.8:

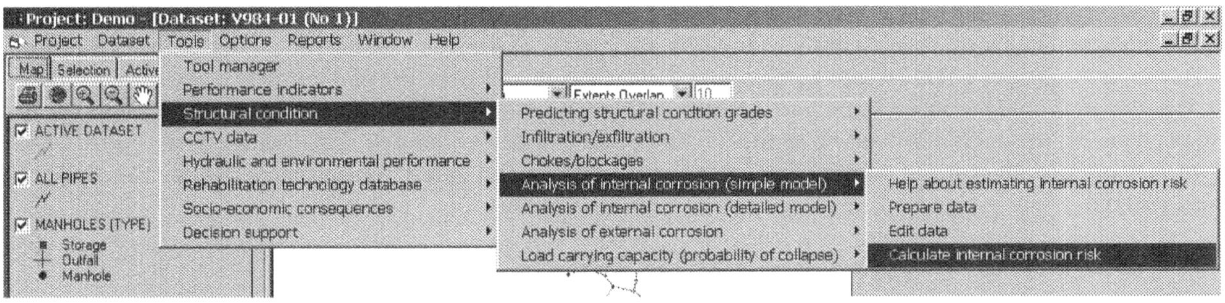

Figure 7.8 The CARE-S 'tools' menu

At the 'Prepare data' request, the CARE-S software creates the necessary input file(s) for an analysis tool from data held in the project database according to a pre-defined specification (managed by the internal Specification Manager). These files may then be edited or used to run the tool within (or outside) the CARE-S framework. The software will then automatically register the results of the tool analysis in to CARE-S, and store them in an appropriate form in the database.

The tools on the 'Tools' menu are grouped according to their analysis function, namely: performance indicators, structural condition, CCTV data, hydraulic & environmental performance, rehabilitation technology database, socio-economic consequences and decision support. The specific tools are listed under these functions and, for each tool, there are a number of generic functions, namely:

- Accessing the brief help of the tool;
- Generating input files from data in the project database (in order that the tool may be used to analyse the active dataset);
- Editing the newly created data files prior to use by the tool;
- Running the tool.

The Expression Builder

Whilst running the various CARE-S tools, users will need to find information, create, select and modify datasets, and submit these datasets to the tools for analysis. These facilities are provided by the Rehabilitation Manager via the GIS toolbar, shown in Figure 7.9.

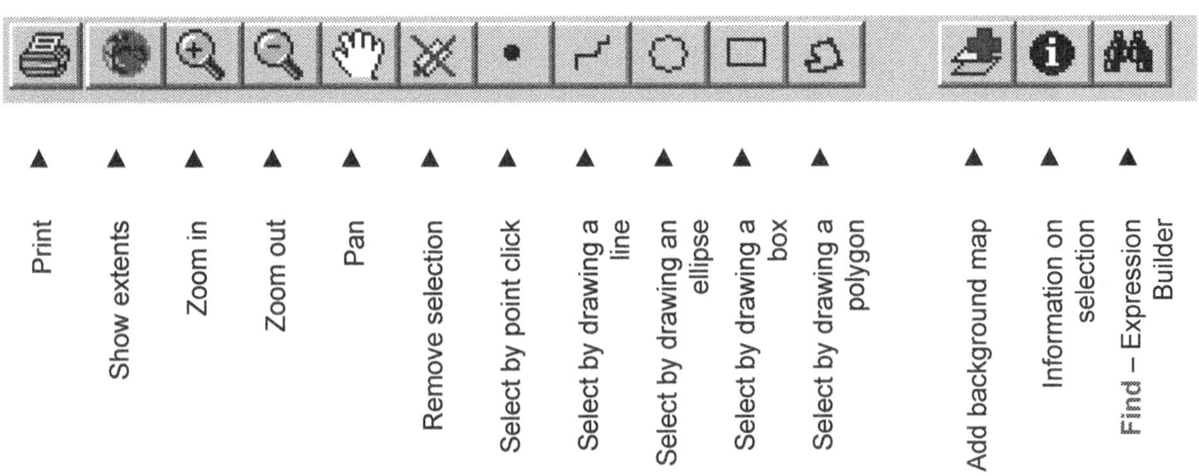

Figure 7.9 The GIS toolbar

CARE-S provides the standard functions found in many GIS applications. Some of these have been modified and some additional features have been included to provide users with functions which are specific to the needs of CARE-S rehabilitation planning. Principal amongst these is the expression builder, which enables users to find and modify information.

The expression builder is accessed via the 'Find' button (see Figure 7.9), which provides a number of pre-defined search queries plus the facility for users to develop and save their own queries. An example of the use of the expression builder is shown in the figures below, where pipes whose diameter is recorded as 152 or 155mm (i.e. converted from 6 inches, in UK imperial units, to metric units) are identified (Figure 7.10) and changed to 150mm (Figure 7.11). The query syntax of the expression builder can be applied to any of the attributes stored in the database; it thus provides a powerful means of searching and modifying the data, and grouping the assets for further analysis according to user-defined criteria.

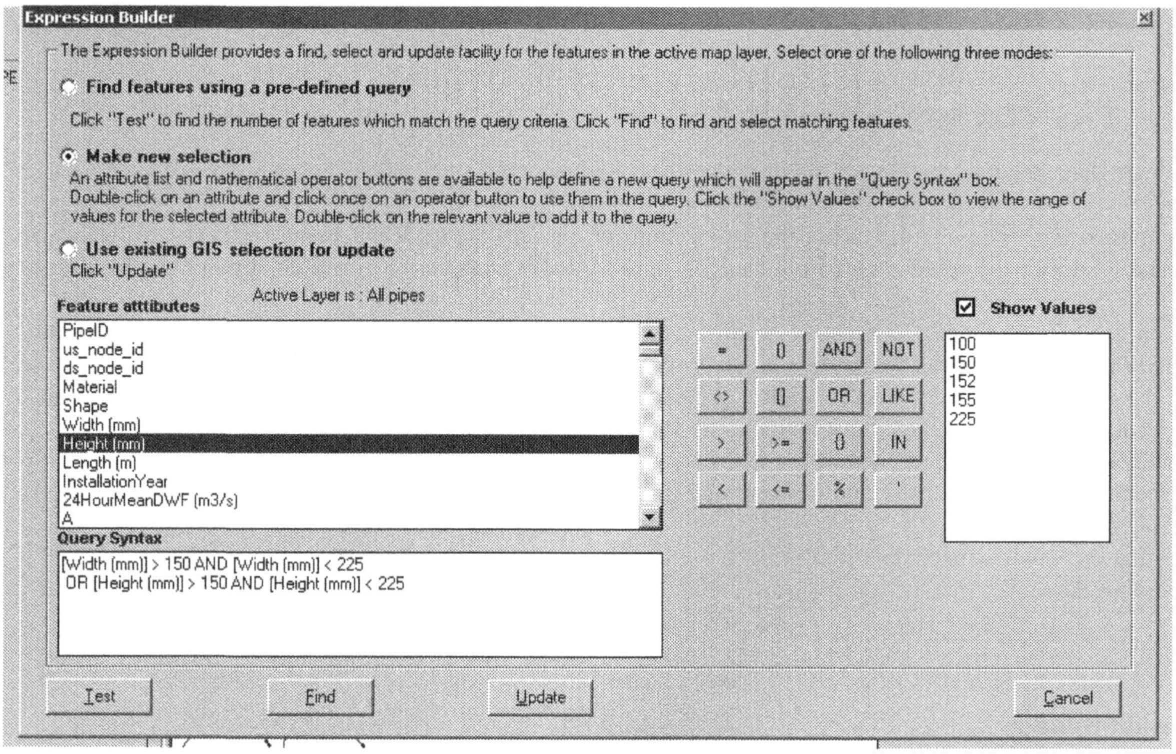

Figure 7.10 Finding assets using the Expression Builder

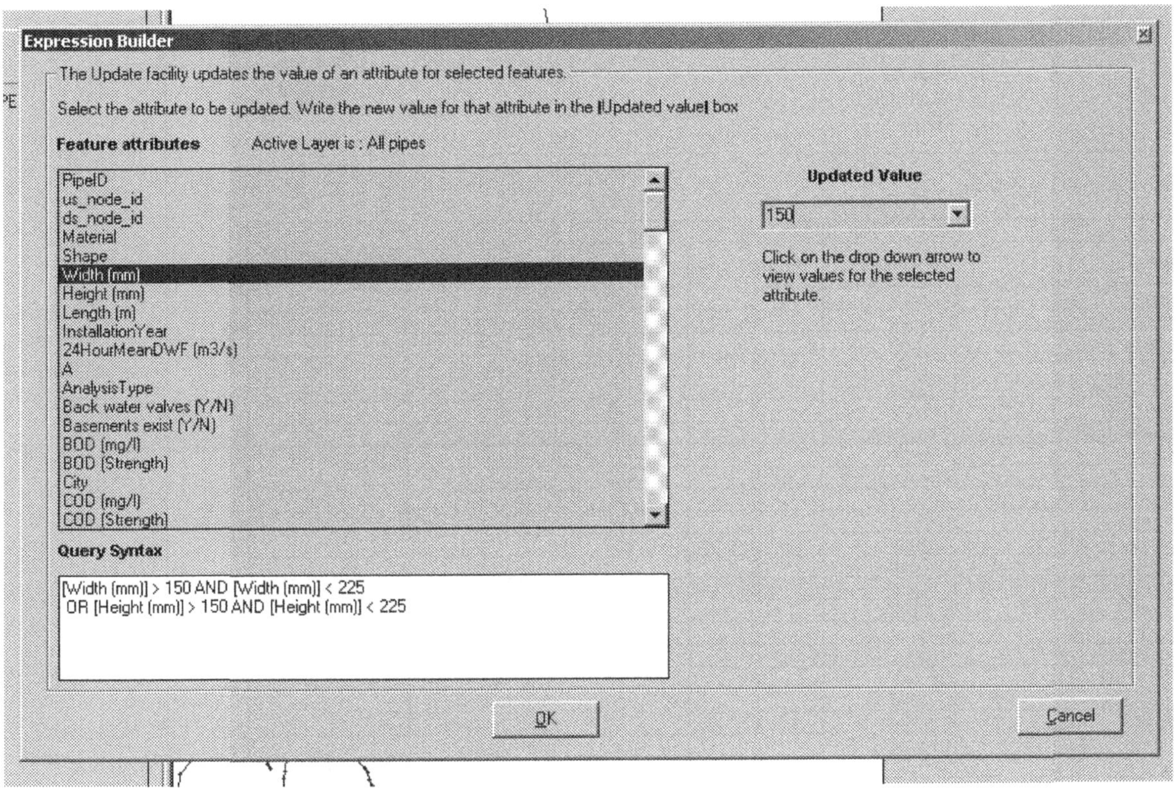

Figure 7.11 Updating asset attributes

Reporting

One of the main objectives of the Rehabilitation Manager is to manage information and results, enabling users to apply the CARE-S tools to identify optimum rehabilitation strategies.

The Rehabilitation Manager can display both project summary information and analysis results. This information can be used to identify gaps in data, assist in the choice of which tools to utilise, aid the selection of sewers to be included in a given analysis and assist in the final development of the rehabilitation plan.

Individual tools also provide reports. The Rehabilitation Manager complements these by enabling results from two or more tools to be combined, providing a more 'global' view of the results. For example, the Rehabilitation Manager can combine results from structural and hydraulic analyses, helping the user to develop integrated solutions to the problems identified.

A set of standard reports is provided, together with the facility for users to specify and generate their own reports. The reports can be produced either in GIS or tabular format. GIS has the obvious advantage of providing the information geographically, which generally assists interpretation. Tabular reports can be exported to other software applications (e.g. MS Excel spreadsheets) for further analysis or presentation as graphs or pie-charts.

Examples of tabular and GIS reports are shown in Figure 7.12 and Figure 7.13.

Asset Summary Report

Dataset name: V984-01
Project: Grottsville
Dataset type: Network

Summary of current status of dataset

Category of pipe				Era of Construction	Total length (km)
Material	Shape	Height (mm)	Width (mm)		
Concrete	Circular	0	300	Pre 1914	0.373
Concrete	Circular	0	375	Pre 1914	0.100
Concrete	Circular	0	450	Pre 1914	0.068
Concrete	Circular	0	525	Pre 1914	0.025
Vitrified clay	Circular	0	150	Pre 1914	0.871
Vitrified clay	Circular	0	225	Pre 1914	0.879
Concrete	Circular	0	300	Inter War	0.322
Concrete	Circular	0	375	Inter War	0.376
Concrete	Circular	0	450	Inter War	0.097
Concrete	Circular	0	600	Inter War	0.111
Concrete	Circular	0	675	Inter War	0.085
Concrete	Circular	0	825	Inter War	0.039
Concrete	Circular	0	900	Inter War	0.078
Concrete	Circular	0	975	Inter War	0.360
Concrete	Circular	0	1125	Inter War	0.219
Concrete	Circular	0	1200	Inter War	0.358
Vitrified clay	Circular	0	150	Inter War	2.003
Vitrified clay	Circular	0	225	Inter War	2.358
Vitrified clay	Circular	0	300	Inter War	0.186
Concrete	Circular	0	150	1945 to 1979	0.349
Concrete	Circular	0	225	1945 to 1979	0.359
Concrete	Circular	0	300	1945 to 1979	1.327
Concrete	Circular	0	375	1945 to 1979	1.758
Concrete	Circular	0	450	1945 to 1979	1.683
Concrete	Circular	0	525	1945 to 1979	0.828
Concrete	Circular	0	600	1945 to 1979	0.438
Concrete	Circular	0	675	1945 to 1979	0.790
Concrete	Circular	0	750	1945 to 1979	0.034
Concrete	Circular	0	825	1945 to 1979	0.730
Concrete	Circular	0	900	1945 to 1979	0.523
Concrete	Circular	0	975	1945 to 1979	0.325
Concrete	Circular	0	1200	1945 to 1979	0.342
GRP	Circular	0	375	1945 to 1979	0.055
GRP	Circular	0	510	1945 to 1979	0.050
GRP	Circular	0	525	1945 to 1979	0.005
PVC	Circular	0	150	1945 to 1979	0.031
Vitrified clay	Circular	0	100	1945 to 1979	0.231
Vitrified clay	Circular	0	150	1945 to 1979	13.643
Vitrified clay	Circular	0	225	1945 to 1979	9.266
Vitrified clay	Circular	0	300	1945 to 1979	1.303
Vitrified clay	Circular	0	375	1945 to 1979	0.028
Vitrified clay	Circular	0	225	Unknown	0.062
Vitrified clay	Circular	0	300	Unknown	0.062

Figure 7.12 CARE-S Asset Summary Report

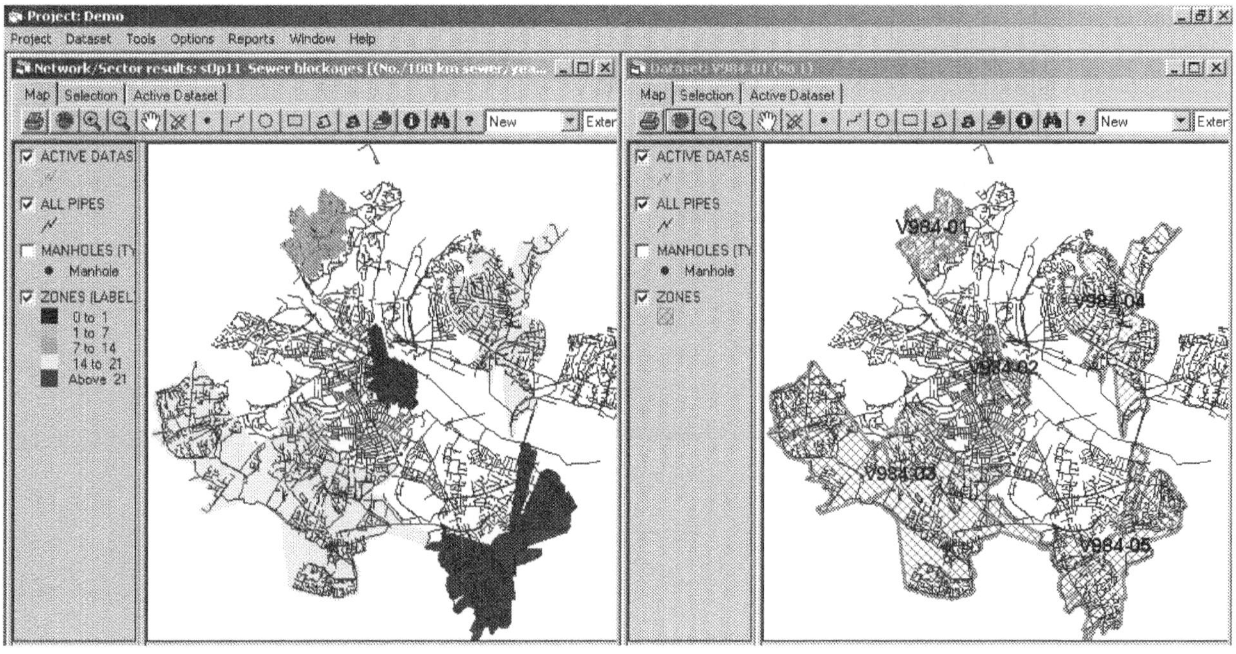

Figure 7.13 Wastewater PI GIS report

Help facilities

Finally, the Rehabilitation Manager provides comprehensive 'Help' facilities grouped under the following headings:

- CARE-S overview
- Glossary of terms
- Quick guide to the tools
- Introduction to Sewerage Rehabilitation Planning
- Software basics
- Working with Projects and Datasets
- Working with CARE-S tools
- Data issues
- Getting results
- Using your results
- Brief help of each of the tools

The Help files provide guidance on the EN752-5 rehabilitation planning procedure and describe how to use the CARE-S Rehabilitation Manager and tools to implement the procedure.

7.4 CONCLUSION - THE ROLE OF THE REHABILITATION MANAGER

The Rehabilitation Manager is an integral part of the CARE-S toolkit. It provides the means of controlling and managing data flows between CARE-S tools and provides the facility for generating datasets for analysis and presenting results.

The Rehabilitation Manager supports, or is used for, the following tasks:

- Providing guidance on the EN752-5 rehabilitation planning procedure;
- Allowing users to implement the EN752-5 procedure by mobilising a range of tools, provided under the CARE-S umbrella;
- Providing a framework by which users can access and interact with the CARE-S tools;
- Selecting pipes that will be analysed by the CARE-S tools;
- Providing a central storage area (the CARE-S database) and reference point for essential data;

- Creating the necessary input file(s) from the data stored in the CARE-S database to enable the tools to be run;
- Accepting and storing necessary output data on completion of tool use;
- Displaying and reporting the results of the analysis;
- Performing rudimentary data validation including duplication, data type and range checking;
- Providing 'Help' facilities which guide the user through the rehabilitation planning procedure and describe how to use CARE-S to implement the procedure.

8

End-user testing

Editors: Rita Ugarelli, Bologna University, Axel Kønig, Sintef

8.1 INTRODUCTION

The testing of the *CARE-S Rehab Manager* (in the proposal the term *Prototype* was used) and all the tools contained within it, was the final part of the *CARE-S* project, but one of the most important. It allowed each software component to be thoroughly evaluated using data from different end-users – data that are diverse in terms of quantity and quality, as well as in geographic origin. It also permitted the whole rehabilitation procedure to be tried, and facilitated the establishment of a recommended best practice, based on data availability and the utility's objectives. This report outlines the testing procedure and presents the results from the testing.

The aim of this WP in the *CARE-S* project was to test the various tools in the *CARE-S* toolkit within the main framework of the *CARE-S* software (*CARE-S Rehab Manager*). In addition to the testing of the integrated *CARE-S Rehab Manager*, the individual tools have also been tested separately. The testing phase addresses both the ease of use of the *CARE-S* software and tools as well as the relevance of the analysis that has been undertaken.

Besides aiding removal of software bugs, such broad testing provides many benefits:

The opportunity to examine data availability in different regions and countries with

- a view to determining the "most appropriate use of *CARE-S*" for each;
- A chance to determine how "*CARE-S* - the *Rehab Manager*" can fit into and improve the existing rehabilitation management of end-users;
- Identification of extra functionality requirements and needs for further research.

The testing phase for the *CARE-S* project officially began at the end of March 2005, version 1.0.0 of the *CARE-S* software was issued. Since then, several versions of the software have been released, being 2.6.0.13 the current one. After finalisation of the project, the final version 3.0.0.0 will be released.

8.2 OVERVIEW OF THE TESTING

This Working Package has been broken down in two tasks as follows:

Task 8.1. System for testing: the toolkit was tested for 8 real cases reflecting different European conditions, including geographical location, size and water/wastewater management organisation.

The testing has been carried out in three steps, according to the *CARE-S Description of Work* document:

1. Identification of the current state-of-art

2. Comparative studies of programmes for pipe condition grades and Service reliability
3. Testing, including evaluation, of the *Rehab Manager*.

Task 8.2. Analysis: A synthesis of the case studies is extracted from the testing to support further business development and implementation of the *CARE-S Rehab Manager*. This synthesis comprises a critical view of the tools and methods and will also include sensitivity and feasibility analyses.

Table 8.1 gives an overview of the test cities, including some properties of water network and management.

The testing phase carried out in this penultimate work package involved all partners and end-users participating within the project. The focus was on testing all tools within the CARE-S Rehab Manager environment rather than as stand alone software. Testing has been performed at several levels by the end-users. Some tools have only been applied towards data transfer and functionality with an artificial data set, whilst others have been analysed more thoroughly, including evaluation of reliability of results. These testing details can be retrieved in the respective test reports in the appendix.

For most of the test cases there is an end-user report providing feedback and suggestions for further improvement of the software. The individual test reports from the end-users are given as appendices of this document.

Table 8.2 Properties of the test zones in the cities

City	**Brno**	**Dresden**	**Oeiras/ Amadora**	**UK**	**Oslo**	**Reggio Emilia**	**Barcelona**	**Palermo**
Country	CZ	D	P	UK	N	I	ES	I
Location in Europe	CE	CE	SW	N	N	S	SW	S
Type of entity	P	M	M	P	M	P	P	M
Tourist area	N	N	Y	N	N	N	Y	Y
Electronic records	Y	Y	Y	Y	Y	Y	Y	Y
CCTV	M	A	M	M	A	M	M	M
GIS	Y	Y	Y	Y	Y	Y	Y	Y
Network length (km)	26	139	4	80	13,7	92,7	3	80
Average pipe age (years)	40	80	25		42	30		10
Hydraulic model	M	S	M	I	M	I		S

Explanation:
1. Location: S,E,N,W,C= south, east, north, west, central
2. Type of entity: P=private, M=municipal
3. Tourist area: Y=yes, N=no, P=partly
4. Electronic records: Y=yes, N=no
5. CCTV: imported M = Manually, A = Automatically
6. GIS: Y=yes, N=no
7. Hydraulic Model: M = MOUSE, I = InfoWorks, S = SWMM

The size of networks (or test zones within the network) varied considerably. In some cases it was possible to use the whole network for one or more of the tools, whilst in other regions, small zones within the network, already identified as problematic, were tested. This variety was important in assessing the suitability of each tool given the number of pipes present. For example, failure prediction tools require as many pipes as possible (i.e. a suitably large sample) to ensure statistical significance, whilst large networks will lead to long computational times for hydraulic reliability models.

Table 8.2 shows which tool groups have been in focus for testing at the various utilities.

Table 8.3. Summary of main testing sites including tools tested

Test Network	*Size of network / main test zone (km)*	*CARE-S PI*	*CARE-S Structural*	*CARE-S Hydraulic*	*CARE-S Socio-economic*	*CARE-S Multicriteria*
Brno, Czech Republic	26	x	x	x		x
Correggio, Italy	92,7	x	x	x	x	x
Dresden, Germany	139			x		
UK, United Utilities	80			x	x	
Oslo, Norway	13,7	x	x	x		
Clabsa, Barcelona	3	x	x	x		
LNEC	3,668	x	x	x		
Palermo	80'	x	x	x	x	x

The following figures give an overview of number of PI calculated by users, CCTV availability and tools tested. The number of PI that have been tested was not given by all end-users even they have applied this tool. Figure 8.1 contents therefore no information of number of applications. The same applies for figure 8.2 where only few end-users gave a percentage of available CCTV data for their network.

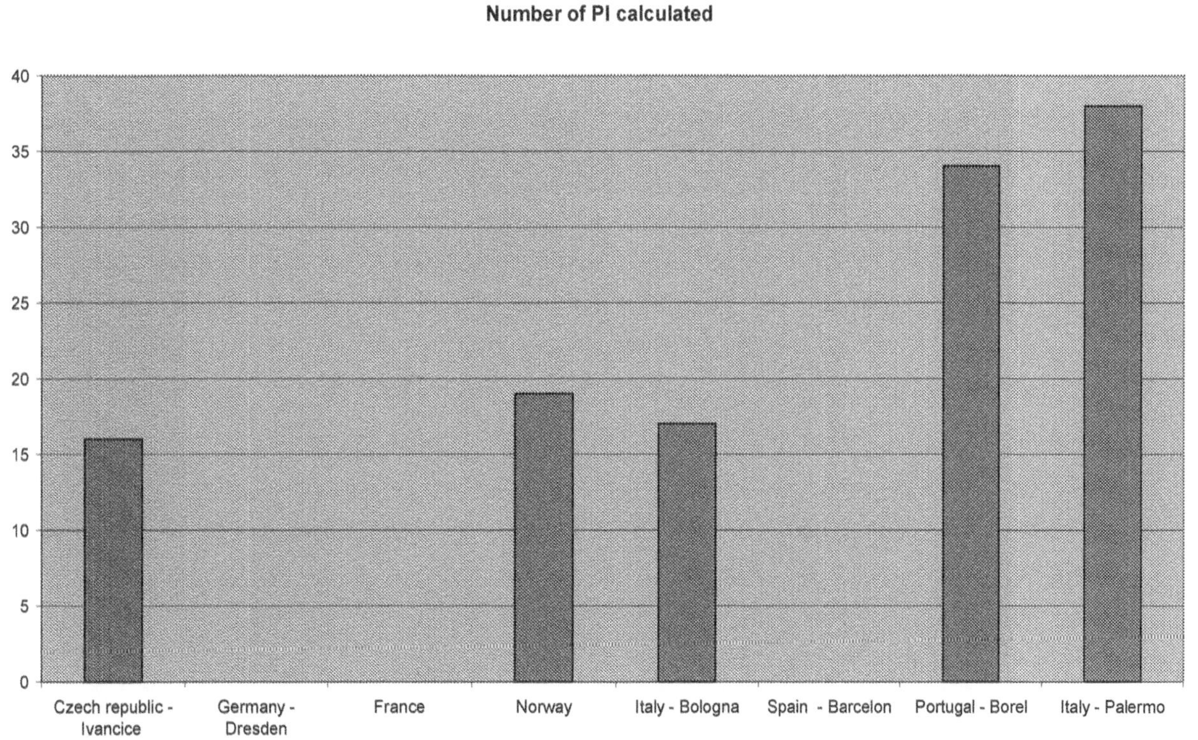

Figure 8.12: Number of PI calculated per user (if known)

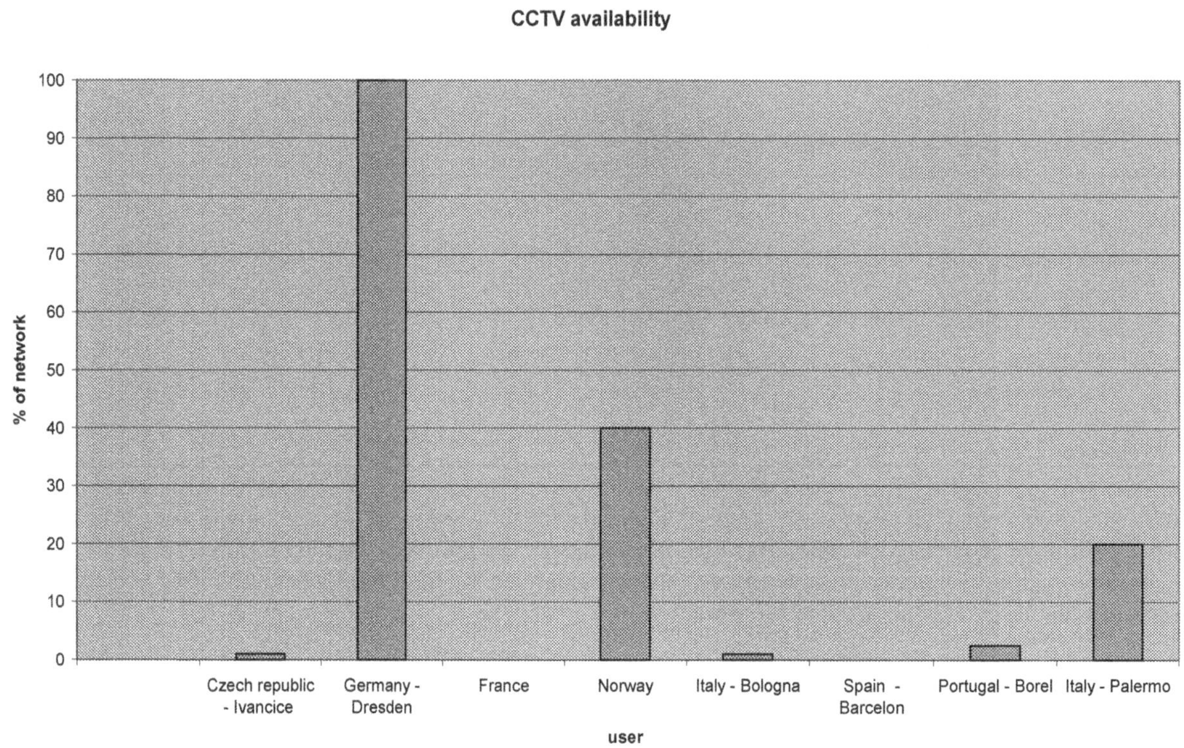

Figure 8.13: CCTV availability in % of network

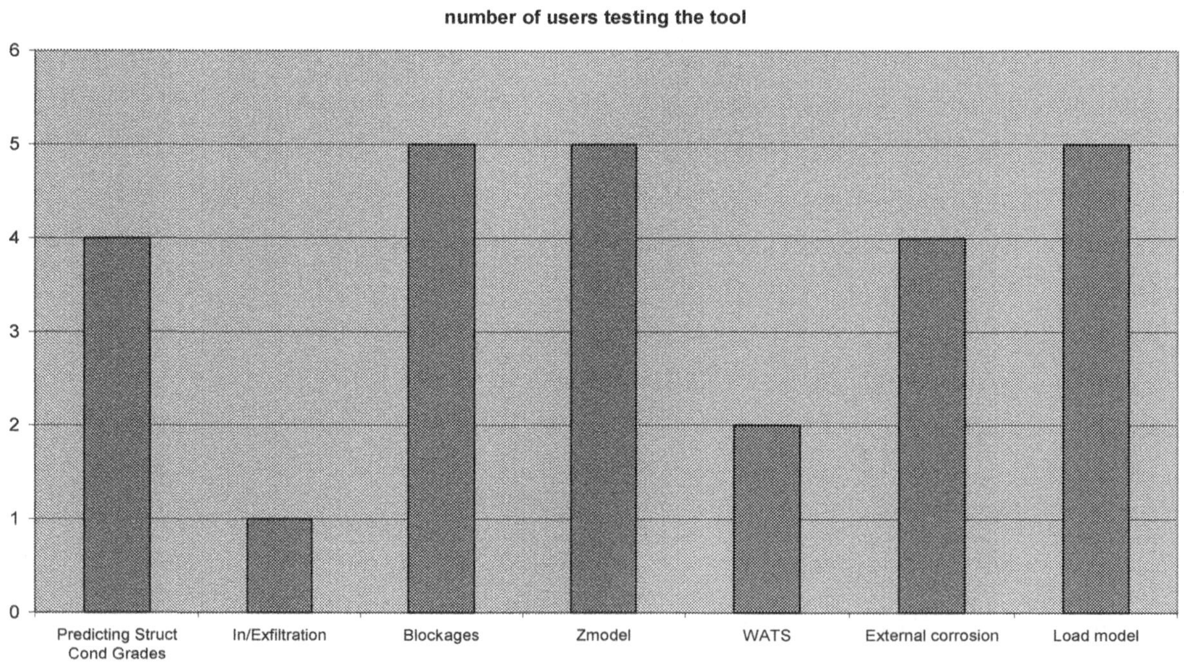

Figure 8.14: Structural tools: number of testing applications

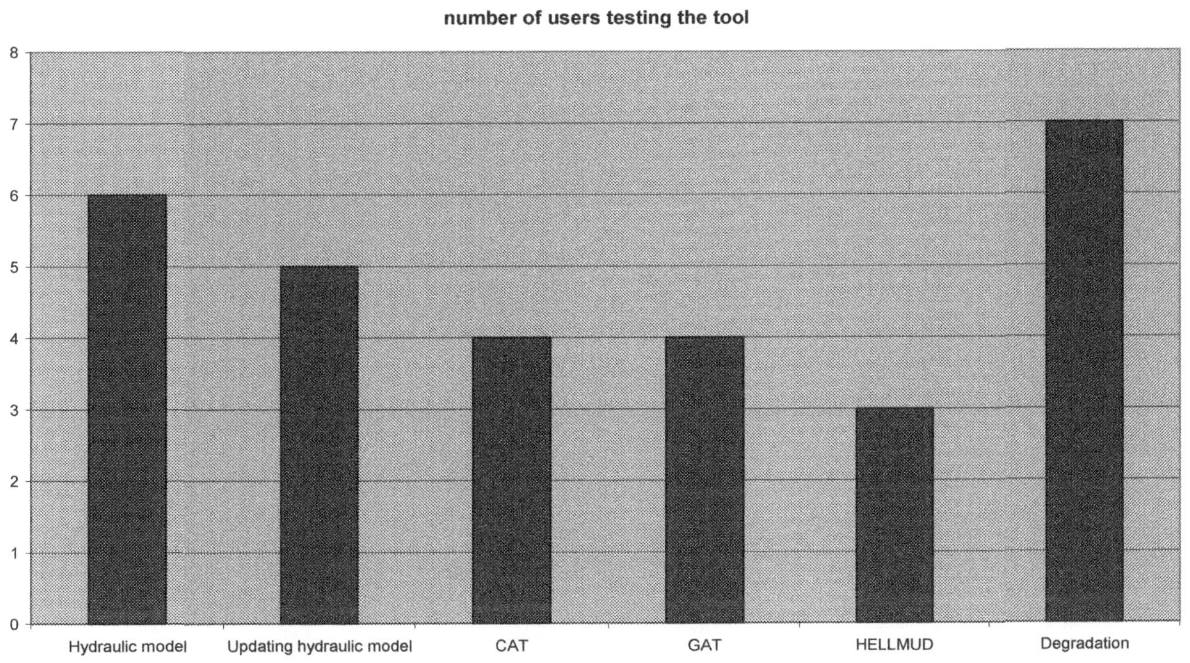

Figure 8.15: Hydraulic Models and tools: number of testing applications

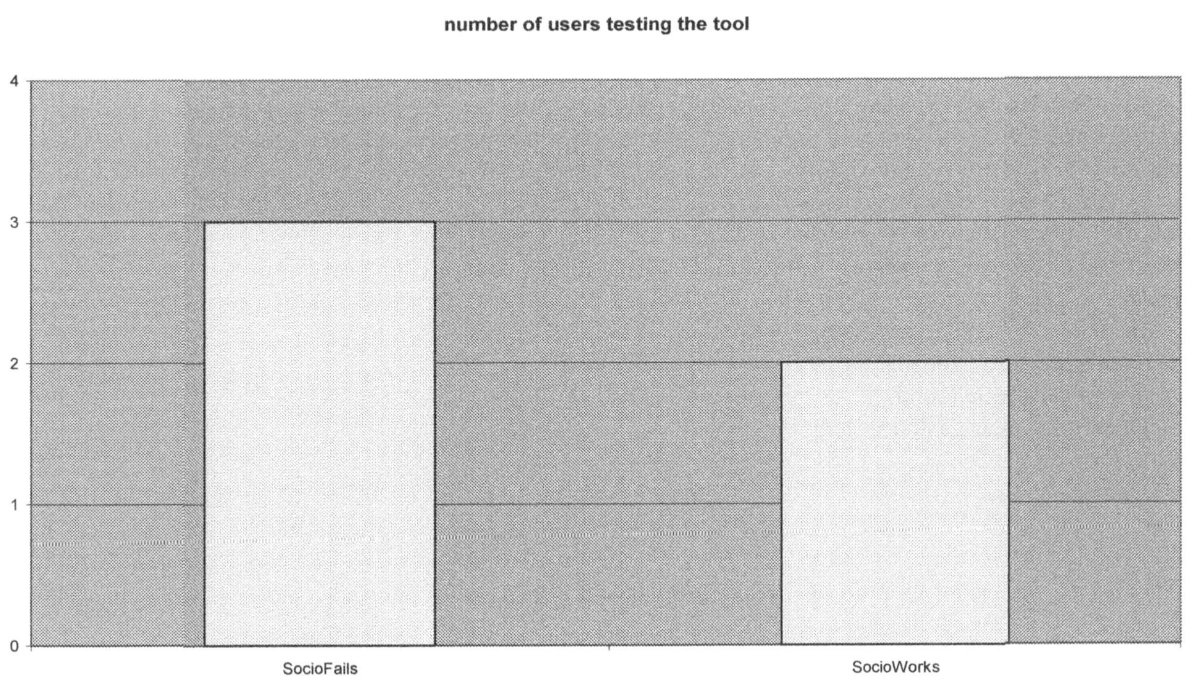

Figure 8.16: Socio – economic tools: number of testing applications

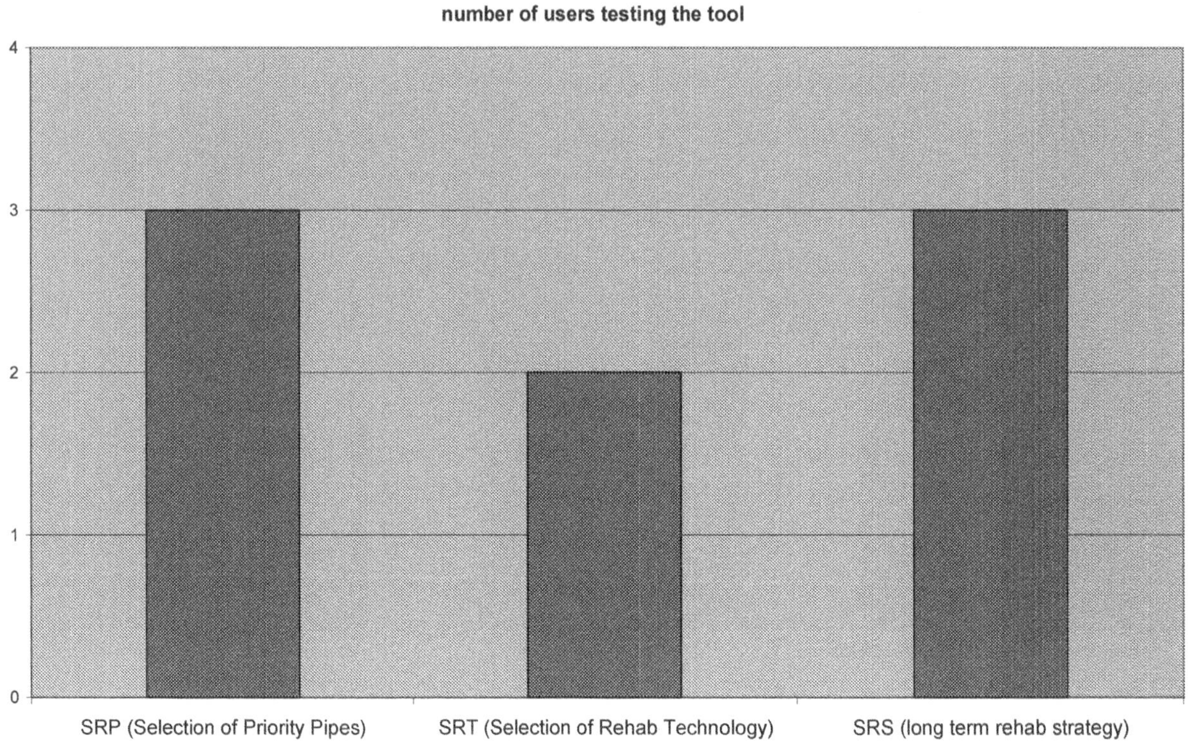

Figure 8.17: Decision support tools: number of testing applications

8.2.1 Testing approach

The testing was carried out by using error/issue log reports. As long as program errors where identified during the testing, partners and end-users were able to report about them. For this purpose, a server was set up with Basic Support for Cooperative Work (BSCW) software in order to provide an internal shared workspace for the project for CARE-s partners. A standard web form was designed and uploaded to the BSCW server intended for that purpose. The users could fill it with all details concerning each error found.

The information required to report any error is:

- Date
- Brief description of issue/error
- Tool where the error first located
- Tool where issue found
- Operating language and system
- Error type
- Contact details

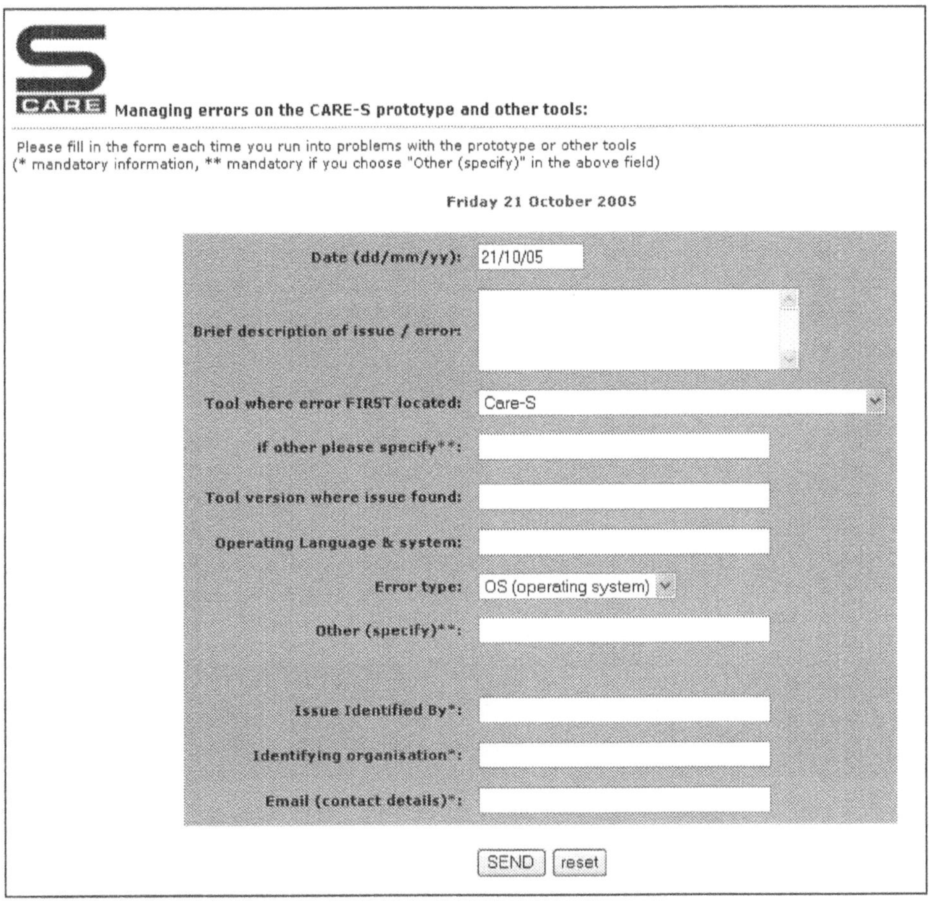

CARE-S Managing errors on the CARE-S prototype and other tools:

Please fill in the form each time you run into problems with the prototype or other tools
(* mandatory information, ** mandatory if you choose "Other (specify)" in the above field)

Friday 21 October 2005

Date (dd/mm/yy): 21/10/05

Brief description of issue / error:

Tool where error FIRST located: Care-S

If other please specify**:

Tool version where issue found:

Operating Language & system:

Error type: OS (operating system)

Other (specify)**:

Issue Identified By*:

Identifying organisation*:

Email (contact details)*:

SEND reset

Figure 18 The CARE-S issues reporting webform

All errors reported were stored in a workbook for further investigation (Figure 7). Once verified and tested the existence of each error, this was reported to the corresponding tool owner with as much details as possible to make possible to reproduce the error and fix it.

8.2.2 Updating tools

The testing process within the wastewater utilities has been performed by the end-user in close collaboration with partners of the CARE-S consortium. End-users could take advantage of the expertise and know-how of technical staff directly involved in the development of the different working packages of CARE-S. This made it possible for the users to validate the usefulness of the software within their utilities. Some utilities were concerned about the feasibility of their own data record-keeping systems to work with CARE-S. This issue also raised awareness of the lack of data (i.e. short periods of records for failure data) that may be needed to start gathering.

On the other hand, partners got feedback from end-users on the usage of the tools and existing bugs which showed up during the testing. In the mean time, the tools were updated accordingly and uploaded to the BSCW server.

8.3 END-USER TEST RESULTS

8.3.1 Data availability and preparation

One of the objectives of *CARE-S* is for the software to be applicable to both data-rich and data-lean networks. Data availability influences the way the *CARE-S Rehab Manager* can be used. Some of the *CARE-S* modules are more data-consuming than others. The advanced user with updated GIS, digital maintenance records, calibrated hydraulic network simulation models, etc. can use all the functionality *CARE-S* provides. However, users with fewer data can also benefit, but will more than likely adopt a different path through the software (Figure 8.8). Exactly which tools to use should be decided case-by-case, based on an evaluation of availability and quality of data.

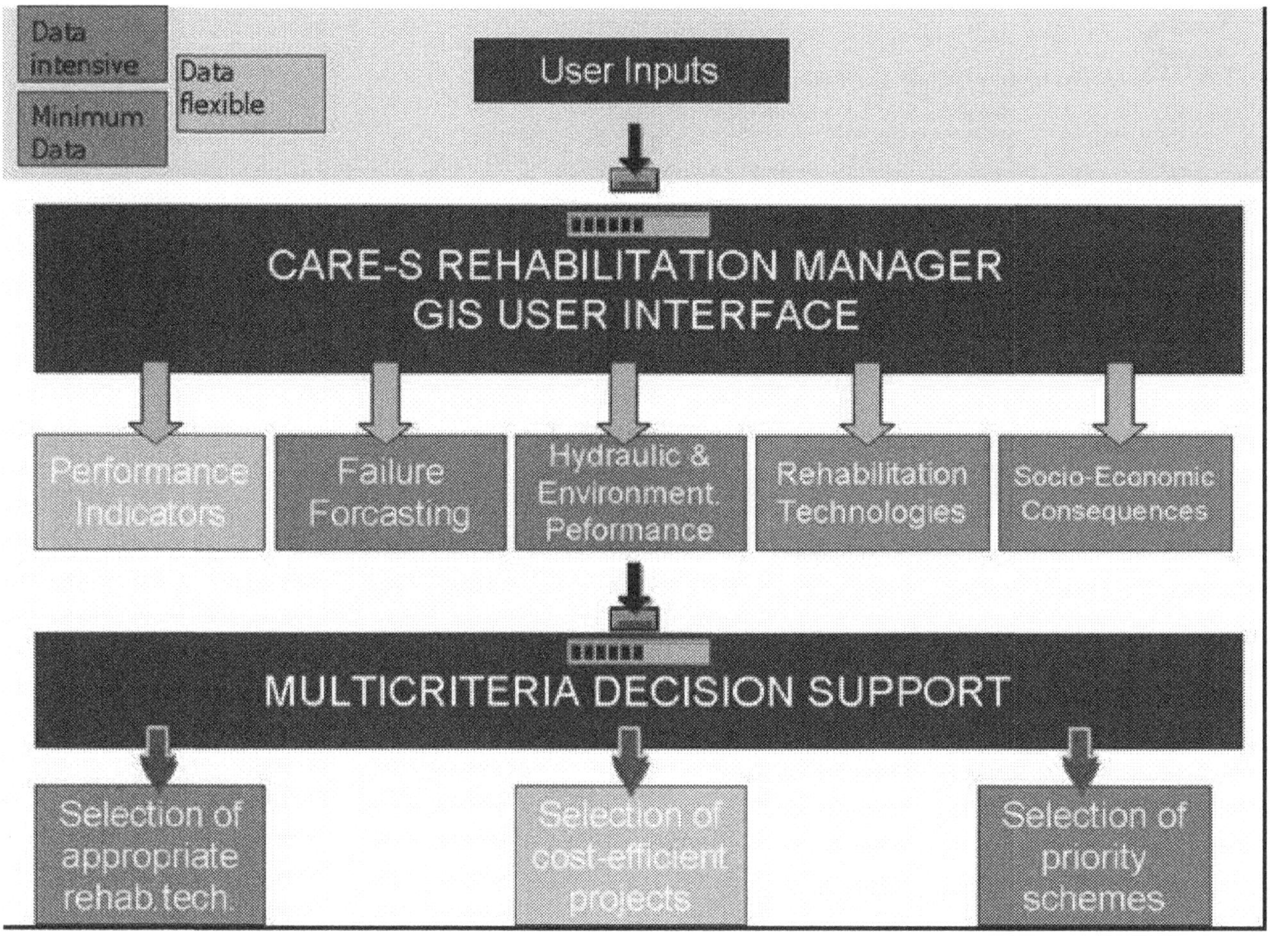

Figure 8.19: CARE-S flow chart and tools data demanding.

CARE-S can also have the role as a catalyst to start or improve the process of recording data and act as an incentive to store as much data as possible of the highest quality. Tools such as failure forecasting models require CCTV inspection records and the variation in record lengths (number of years of data) amongst the test sites, helps determine a minimum recommended value.

Testing has been done by end-users with various versions of the tools during the last year in the project. The results might therefore not always be consistent regarding documentation of problems. Testing has proved to be very valuable in terms of tool improvement and reported problems and bugs have been solved by the respective tool developer in the course of the testing period. Additionally, application of *CARE-S* to real cases unmasked obstacles existing on corporate databases to implement the software: lack of data, quality, etc. At the end of the CARE-S project most tools obtained a satisfying functionality as outlined in Table 8.3.

Some of the listed tools as SocioFails, SocioWorks and SRP, SRT work as stand alone tools, but since their input data are directly prepared by the CARE-S prototype after selection of pipes candidate for rehabilitation plan, the use of that tools as stand alone is not really meaningful.

In general, the preparation of raw data to supply the software in order to be able to run is still an important aspect of the *CARE-S* and represents the major workload of *CARE-S* projects. On the other hand, this is maybe the most critical part of the project. In fact, results from *CARE-S* will depend on the quality of the input data: any analysis undertaken using the *CARE-S Rehab Manager* will only be as good as the data used. Therefore, in order to produce reliable results the input data has to be checked.

The *Rehab Manager* in its present version does not supply features for automatic quality control of ALL the input data. It is the user's responsibility to ensure that valid data are available for import into the *Rehab Manager*.

Table 8.4: State of the art of the CARE-S tools

Tool	Still Bugs	Ok as stand alone	Ok inside prototype
PI	-	Yes	Yes
CCTV Converter	-	Yes	Yes
Predicting Struc Cond grades	Yes	Yes	-
Blockages	-	Yes	Yes
Zmodel	-	Yes	Yes
WATZ	-	Yes	Yes
External corrosion	-	Yes	Yes
In - exfiltration	-	Yes	Yes
Load Model	-	Yes	Yes
Degradation	-	Yes	Yes
CAT	Yes	Yes	-
GAT	-	Yes	Yes
HELLMUD	-	Yes	Yes
Rehabilitation Tech Database	-	Yes	Yes
SocioFails	-	Yes	Yes
SocioWorks	-	Yes	Yes
SRP	-	Yes	Yes
SRT	-	Yes	Yes
SRS	-	Yes	

8.3.2 Main findings

In general, no problems were encountered during the installation of the individual tools. The installation procedure of the *CARE-S Rehab Manager* was not found however particularly straightforward. However, installing subsequent upgrades is much easier and there is no need to remove and reinstall the software longer.

There is a wide variety of database structures being currently used by utilities.

Moreover, structural and maintenance data use to be stored in separate databases. In the *CARE-S Rehab Manager*, GIS asset and failure data are imported as shape files (**.shp*) or comma separated value (‘*.csv*). Some end-users use GIS software which does not produce shape files, being therefore necessary to extract data with other external tools in order to produce shape files importable by the *CARE-S Rehab Manager*. Nonetheless, several features like selection of elements in the GIS interface, expression builder and help file have proved usefulness.

Since not every tool could be tested completely by all end-users, a procedure was performed in order to evaluate the applicability of each tool to their particular situation.

CARE-S_PI

Regarding the individual tools, the *CARE-S_PI Tool* allowed to retrieve a complete outline of the network performance and conditions, disclosing needs for rehabilitation activities.

PI availability on utilities highly depends on recording habits. End-user have their own data recording system which may allow for the calculation of some indicators in the *CARE-S_PI Tool*, It proved to be useful for finding network zones with potential for rehabilitation and to monitor its effects. The *CARE-S_PI Tool*, together with the built-in GIS system on the *CARE-S Rehab Manager* have provided end-user with suggestion on the methodology to collect data.

Structural Analysis(CCTV Converter, Predicting Structural Condition grades (GompitZ), in-exfiltration, Blockages, Zmodel, WATZ, External corrosion, Load Model)

Setup and run of *CARE-S_Structural tools* encountered no problems and was found to be easy and straightforward. The user-friendly interface made possible an intuitive usage of most of the listed tool. However the need of CCTV inspection data available for condition grade prediction limited the use and testing of the GompitZ tool and tools related with its results (SRS, SRP). The GompitZ tool is not completely implemented in the prototype, the need of running it as stand alone in a DOS window reduces the possibilities of application of the tool.

Hydraulic Analysis (Degradation, CAT, GAT, HELLMUD)

Setup and run of *CARE-S_Hydraulic tools* encountered no problems and was found to be easy and straightforward. The user-friendly interface made possible an intuitive usage of most of the listed tool.

Problems incurred with the use of the COSs Assessment Tool (CAT): the tool is not implemented yet in the prototype, but can run as stand alone.

Rehabilitation Tech Database

The Rehabilitation Technology database has been positively evaluated by users. The possibility of adding their specific technology names and select only the one really applied by the utility represents a useful feature. The tool is considered useful also as stand alone database.

Socio-economic Analysis (SocioFails, SocioWorks)

Socio and economic tools have been tested in few networks. The two tools developed are highly data hungry and their application doesn't appear easy to understand for users.

Informatic improvement should be added in order to make the tools

Decision Support (SRP, SRT, SRS)

Setup and run of *CARE-S_SRP and SRT tools* encountered no problems and was found to be easy and straightforward. The user-friendly interface made possible an intuitive usage of most of the listed tool. However, for SRP, special care had to be taken when planning rehabilitation policies: if socio-economical issues are taken into account as criteria for annual rehabilitation plans, results might significantly differ from those obtained if only hard facts such hydraulic reliability or condition grades are considered.

SRS tool needs informatics development in order to completely implement it in the CARE-S prototype and to allow a more user friendly interaction with the GompitZ tool, only manually done so far.

CARE-S usage by the end-user might be a tough task to undertake if there is no additional consultation supports form the partners. This is an important finding highlighted several times on this testing: proper training is therefore essential.

8.3.3 Updated help files

Each of the individual tools, and also the *CARE-S Rehab Manager* has its own help system.

The testing proved the need of end-users for a suitable on-line help system that makes possible an easy usage of the software and a better understanding of the scientific fundamentals behind each individual tool.

The Help file has been made using "HTML HELP Control Version 5.2.3750.1280".

The Help files created during the first stages of development of the tools have been kept updated according to the suggestions made by end-users and when new features have been added to the software.

The quality on the Help system varies between the different tools. Some are more clear or user-friendly than others.

A common Help file is available inside the CARE-S prototype.

The user can browse the different help topics in different ways: through navigation buttons or by means of index of contents, or searching facilities.

The CARE-S Help file guides the user through the topics:

- CARE-S overview
- Software basics
- Data issues
- Working with CARE-S tools
- Working with projects and datasets
- Getting results
- Using your results
- Brief help of the tools

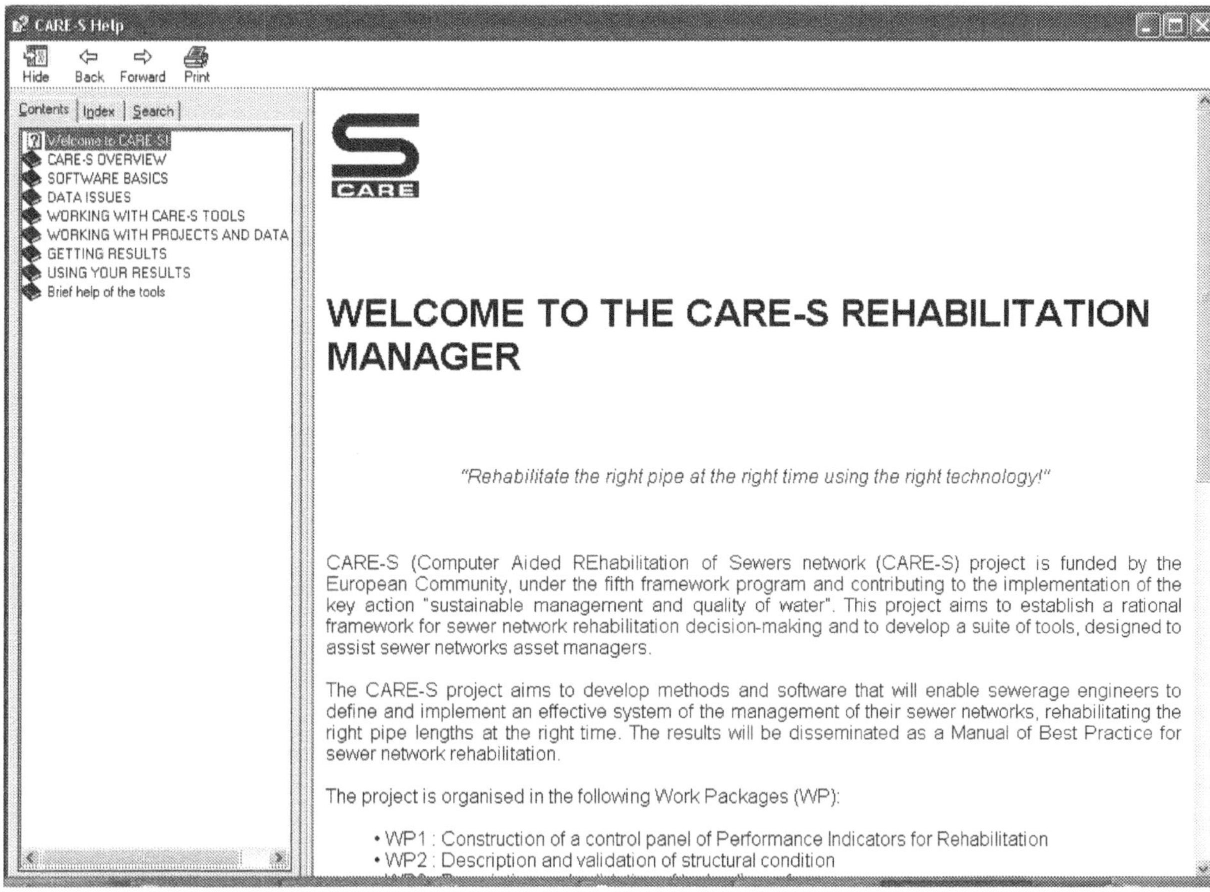

Figure 8.20: CARE-S Help File

8.3.4 Proposals for enhancement

The maintenance and upgrading of the *CARE-S* software and toolkit is an issue that needs to be addressed in the commercial phase of *CARE-S* .

Hydraulic results are provided by external software. As long as *CARE-S* results rely to certain extent on that software, compatibility should be maintained for future versions of this.

The CARE-S prototype needs to be improved in order to include hydraulic simulations run using the freeware model SWMM.

Some tools, e.g. *CAT, GompitZ, SocioFails/Works*, still require much computer resources to run in a user friendly way.

Limits have been highlighted by users in CCTV availability: users often suffer of lack of data on pipe inspected. In many cases the percentage on inspected pipes is around 1% and, when CCTV are available, they are not digitalised but recorder on written protocols.

Care-S, as Care-W, are data hungry and time demanding on data collection.

Another weakness is that a very small number of end-user performs hydraulic modelling; and, in turn, the lack of hydraulic data limits the use of Care-S.

As a consequence, an end user desiring to use Care-S must be prepared to invest a large amount of money and time in collecting data and performing hydraulic modeling.

CARE-S usage by the end-user might be a tough task to undertake if there is no additional consultation supports form the partners. This is an important finding highlighted several times on this testing: proper training is therefore essential.

Other minor improvements, although not critical, might be considered to be include within the software. Better integration with the Windows environment can be achieved by using controls from the Microsoft *Common Dialogs* dynamic link library in many of the forms within the tools. This is applicable to most of the tools so there is still some room for the improvement of the interface, reducing the opportunity for user error, providing all the tools of *CARE-S* with a similar appearance.

Some users appreciated also the possibility to have *CARE-S* translated to the local language.

8.4 CONCLUDING REMARKS

Overall, the tools included in CARE-S can work successfully with the CARE-S Rehab Manager. All bugs identified during the testing have been fixed and validated. Compatibility between input and output data from CARE-S Rehab Manager and the individual tools has been found to be an important issue, and proved to be acceptable at the current stage. Compatibility with existing information systems used by utilities is still an issue to be addressed.

The testing carried out by end-users made it possible to evaluate data availability on corporate databases as well as their structure, thus enabling the way for future implementation of rehab programmes based on CARE-S. New ideas for data recording have been identified by the end-users. On the other hand, extra functionality needed in the software was identified.

Feedback provided by the testing and validation of CARE-S extends and enhances the current possibilities to approach rehabilitation of Sewer networks. End-users participating in the testing of CARE-S generally validated the software as a useful tool which fits with those already being used in water/wastewater utilities for asset maintenance management. It is expected that the CARE-S toolkit as a whole will allow for the definition of a recommended best practices for future rehabilitation plans, based on data availability and management goals.

IWA Publishing's authorised EU representative for General Product Safety Regulations is Diane D'Arras, 15 rue Duret, 75116 Paris, France, e-mail: safety@iwap.co.uk.

Printed and bound by CPI Group (UK) Ltd, Croydon, CR0 4YY
12/05/2026
02108892-0019